THE INTELLECTUAL PROPERTY AND FOOD PROJECT

The Intellectual Property and Food Project

From Rewarding Innovation and Creation to Feeding the World

Edited by

CHARLES LAWSON and JAY SANDERSON
Griffith Law School, Australia

Routledge
Taylor & Francis Group
LONDON AND NEW YORK

First published 2013 by Ashgate Publishing

Published 2016 by Routledge
2 Park Square, Milton Park, Abingdon, Oxon OX14 4RN
711 Third Avenue, New York, NY 10017, USA

Routledge is an imprint of the Taylor & Francis Group, an informa business

British Library Cataloguing in Publication Data
A catalogue record for this book is available from the British Library

The Library of Congress has cataloged the printed edition as follows:
The intellectual property and food project : from rewarding innovation and creation to feeding the world / By Charles Lawson and Jay Sanderson.
p. cm.
Includes bibliographical references and index.
ISBN 978-1-4094-6956-8 (hardback)
1. Plants, Cultivated—Patents. 2. Biotechnology—Patents. I. Lawson, Charles, editor of compilation. II. Sanderson, Jay (Law teacher) editor of compilation.

K3876.I577 2013
346.04'86—dc23

2013020863

ISBN 9781409469568 (hbk)

Contents

List of Tables, Figures and Boxes

Tables

Figures

Boxes

Notes on Contributors

Berris Charnley is a Research Fellow with the Australian Centre for Intellectual Property in Agriculture at Griffith University. He helped establish the IPBio Network as an international collection of scholars with a shared research interest in the history of IP in the biosciences and its relevance to current policy. Berris researches the history and philosophy of science (and in particular biology, plant breeding, genetics and genomics), the history and theory of IP, and the history of food science.

Graham Dutfield is the Professor of International Governance at the University of Leeds. He is also a Research Affiliate of the Intellectual Property Law and Technology Program at Osgoode Hall Law School, York University, Toronto, an Adjunct Professor at the Center for Studies of Intellectual Property Rights at Zhongnan University of Economics and Law, Wuhan, China, a member of the IPBio Network and a member of the Scientific Advisory Board of the Canadian synthetic biology project PhytoMetaSyn. Graham researches IP across several disciplines, including law, history, politics, economics and anthropology.

Robert J. Henry is a Professor at the University of Queensland and the director of the Queensland Alliance for Agriculture and Food Innovation. He has been involved with the Grains Research and Development Corporation, Sugar Research and Development Corporation and Rural Industries Research and Development Corporation committees, and was a member of the Australian Plant Breeders' Rights Advisory Committee. Robert researches improved food and energy security by applying biochemical and molecular tools to the development of improved crop varieties, analysing domesticated crops, wild relatives of crop species and potential new crop species.

Charles Lawson is an Associate Professor at Griffith University and a member of the Australian Centre for Intellectual Property in Agriculture. Charles researches IP, competition, trade, access and benefit-sharing, and public administration.

Karinne Ludlow is a Senior Lecturer at Monash University. Karinne recently co-authored a report for the United Nations Food and Agriculture Organisation on nanotechnology and food – *Nanotechnology Applications in Food and Agriculture Sectors: Principles and Guidance for Food Safety Regulation*. Karinne researches interdisciplinary problems of science and technology and the law, with particular emphasis on biotechnology and nanotechnology.

Matthew Rimmer is an Australian Research Council Future Fellow, an Associate Professor at the ANU College of Law and an Associate Director of the Australian Centre for Intellectual Property in Agriculture. Matthew researches IP and climate change, and has broader interests in copyright law and information technology, patent law and biotechnology, access to medicines, clean technologies and traditional knowledge.

Jay Sanderson is a Lecturer at Griffith University and a member of the Australian Centre for Intellectual Property in Agriculture. Jay researches IP with an emphasis on the relations between plants, science, politics, people and practice.

Brad Sherman is a Professor of Law in the Griffith Law School at Griffith University and Director of the Australian Centre for Intellectual Property in Agriculture. He has been involved with the Australian Centre for International Agricultural Research, the Grains Research and Development Corporation, the Cotton Research and Development Corporation and Rural Industries Research and Development Corporation, and was a member of the Australian Plant Breeders' Rights Advisory Committee. Brad researches intellectual property broadly including international copyright law and practice, copyright law in multimedia, artistic works and public performance, the historical and conceptual origins of modern patent law, patent law in biotechnology, public sector research and technology, and plant intellectual property law.

Brendan Tobin is a Research Fellow with the Australian Centre for Intellectual Property in Agriculture at Griffith University. He has a long record of involvement in international environmental law and protection of the rights of indigenous peoples and local communities. Brendan researches natural resource management, access to genetic resources and benefit sharing, protection of rights over traditional knowledge, and indigenous peoples' human rights.

Acknowledgements

The IP and food project is an ongoing analysis of the ways that IP affect the various dimensions of food as it is broadly conceived. This is of interest because it is a practical application of the ways in which IP actually works, and provides an understanding of the ways that IP affect food from the very early stages of research and development through to consumption of products and services by consumers. According to this frame, IP can have both positive and negative impacts. The IP and food project is an ongoing and long-term initiative, and this collection is a step along that road. Over time the Australian Centre for Intellectual Property in Agriculture and the Australian Research Council's *Discovery Projects* funding scheme (DP120101434) have generously supported this project. And our colleagues have contributed through discussions, insights and the assistance that makes the practical outcomes happen. We thank, in particular, Carol Ballard and Julia Barker who facilitated most of the behind-the-pages work.

Abbreviations

ACIPA	Australian Centre for Intellectual Property in Agriculture
ACLU	American Civil Liberties Union
ACP	African, Caribbean and Pacific Group of States
AHTEG	Ad Hoc Technical Expert Group
AnGR	Animal genetic resources
ASSINSEL	Association Internationale des Selectionneurs pour la Protection des Obtentions Vegetales
ASTA	Agricultural Seed Traders Association
CATCH	Choiseul Association of Craft and Heritage Tourism
CBD	*Convention on Biological Diversity*
CIFOR	Center for International Forestry Research
CGIAR	Consultative Group on International Agricultural Research
CGRFA	Commission on Genetic Resources for Food and Agriculture
CIAT	International Center for Tropical Agriculture (Centro Internacional de Agricultura Tropical)
CIMMYT	Centro Internacional de Mejoramiento de Maiz y Trigo
CIP	Centro Internacional de la Papa
COAG	Council of Australian Governments
COGEM	Netherlands Commission on Genetic Modification
CoOL	Country-of-origin labelling
CPB	*Cartagena Protocol on Biodiversity*
DNA	Deoxyribonucleic acid
EC	European Commission
EPA	*Economic Partnership Agreement with the European Union*
EU	European Union
FAO	Food and Agriculture Organization of the United Nations
FSANZ	Food Standards Australia New Zealand
GIs	Geographical indications
GM	Genetically modified
GMO	Genetically modified organism
GPLPGR	General public license for genetic resource
GTR	Gene Technology Regulator
IAASTD	International Assessment of Agricultural Knowledge, Science and Technology for Development
IARC	International Agricultural Research Centers
IBPGR	International Board for Plant Genetic Resources
ICARDA	International Center for Agricultural Research in the Dry Areas

ICESCR	*International Covenant on Economic Social and Cultural Rights*
ICLARM	Worldfish
ICRAF	World Agroforestry Centre
ICRCGC	International Climate Resilient Crops Genomics Consortium
ICRISAT	International Crops Research Institute for the Semi-Arid Tropics
IFPRI	International Food Policy Research Institute
IITA	International Institute of Tropical Agriculture
ILO	International Labour Organization
ILRI	International Livestock Research Institute
INAO	Institut National de l'Origine et de la Qualité
INIBAP	International Network for the Improvement of Banana and Plantain
IPGRI	International Plant Genetic Resources Institute (Bioversity International)
IP	Intellectual property
IRD	Institut de Recherché pour le Développement's
IRRI	International Rice Research Institute
ISNAR	International Service for National Agricultural Research
IU	FAO's *International Undertaking on Plant Genetic Resources for Food and Agriculture*
IUCN	International Union for the Conservation of Nature
IWMI	International Water Management Institute
LMOs	Living modified organisms
MORT	Machine or transformation test
MTA	Material Transfer Agreement
NAFTA	North American Free Trade Agreement
NGOs	Non-governmental organizations
NIAB	National Institute of Agricultural Botany
OECD	Organization for Economic Co-operation and Development
PBR	Plant Breeders' Rights
PCR	Polymerase chain reaction
PDOs	Protected designations of origin
PGIs	Protected geographical indications
PIPRA	Public Intellectual Property Resource for Agriculture
Plant Treaty	*International Treaty on Plant Genetic Resources for Food and Agriculture*
PMSEIC	Prime Minister's Science, Engineering and Innovation Council
PPA	US Plant Patent Act of 1930
R&D	Research and development
RHS	Royal Horticultural Society
RUTF	Ready-to-use therapeutic food
SECs	Socio-economic considerations
SGRP	System-wide Genetic Resources Programme
SISAN	National Food and Nutritional Security System
SMTA	Standard Material Transfer Agreement

SPS Agreement	*Agreement on the Application of Sanitary and Phytosanitary Measures*
TAC	Technical Advisory Committee
TBT Agreement	*Agreement on Technical Barriers to Trade*
TRIPS	*Agreement on Trade-related Aspects of Intellectual Property Rights*
TRRs	Traditional Resource Rights
TSGs	Traditional specialities guaranteed
UN	United Nations
UNDP	United Nations Development Programme
UNDRIP	*Declaration on the Rights of Indigenous Peoples*
UNEP DGEF	United Nations Environment Programme's Division of Global Environment Facility Coordination
UNICEF	United Nations Children's Fund
UPOV	International Union for the Protection of New Varieties of Plants
US	United States of America
USPTO	United States Patent and Trademark Office
VMH	Victoria Medal of Honour
WARDA	Africa Rice Center
WFP	World Food Programme
WIEWS	World Information and Early Warning System of Plant Genetic Resources for Food and Agriculture
WIPO	World Intellectual Property Organization
WTO	World Trade Organization

Chapter 1

The IP and Food Project

Jay Sanderson and Charles Lawson

There is debate about the role intellectual property (IP) law plays in food production and availability. Most broadly, IP is either seen to encourage innovation in food and food-related developments, or, alternatively, to hinder and restrict access to, and exchange of, essential materials and processes such as plant genetic resources. More specifically, though, IP and food interact in innumerable ways – from providing protection to plant and animal breeders, protecting distinctive food brands and logos, helping to identify goods that originate from a particular region or influencing the publication and reproduction of information and data generated through agricultural, scientific and technological research. And while the relationship between IP and food can sometimes be tenuous, the relationship is influential: affecting the production and availability of food by regulating dealings in products, processes, innovations, information, data, and so on.

The scope and nature of the interaction between food and IP is influenced by international laws that include the Convention of the International Union for the Protection of New Varieties of Plants (UPOV) and the *Agreement on Trade-related Aspects of Intellectual Property* (TRIPS). The TRIPS agreement requires all World Trade Organization (WTO) members to adhere to minimum standards of IP as well as observe the *Paris Convention on Industrial Property*, the *Berne Convention on Copyright* and other WTO conventions. In relation to food, TRIPS requires WTO members to provide patent protection for all inventions, whether of products of processes, in almost all fields of technology. While IP protection is optional for plants and animals (other than micro-organisms), as well as for other essentially biological processes used in the production of plants and animals (other than microbiological processes), members of the WTO must provide protection of plant varieties either by patents, or by an effective *sui generis* system, or any combination of patents or plant variety protection. In addition to patents and plant variety protection the TRIPS agreement also sets minimum standards in relation to trademarks, copyright, trade secrets and geographical indications. In addition to specific international laws on IP, the scope and nature of the interaction between food and IP is made more complicated by international treaties that deal with the conservation of biodiversity, as well as access and exchange of plant genetic resources related to food and agriculture. These treaties include the *Convention on Biological Diversity* (CBD) and the *International Treaty on Plant Genetic Resources for Food and Agriculture* (*Plant Treaty*).

Importantly, IP has also found its way into organizations and forums that it has traditionally been absent from. Take, for example, the introduction of mandatory IP principles in the Consultative Group on International Agricultural Research (CGIAR) system. Established in 1971 to support research and development that improves food security and reduces poverty, CGIAR is a global network of agricultural research Centers 'dedicated to reducing rural poverty, increasing food security, improving human health and nutrition, and ensuring more sustainable management of natural resources'.[1] More specifically, some of the CGIAR Centers concentrate their research efforts on improving major food crops such as rice, wheat, maize and potatoes, as well as livestock and fish. To ensure that the Centers' research activities and outputs are freely and globally available for researchers, plant breeders and farmers, they have been, and continue to be, considered international public goods. In March 2012, however, CGIAR introduced *Principles on the Management of Intellectual Assets* that allow, in certain situations, exclusive use of CGIAR Centers' research including the 'prudent and strategic' use of IP protection.[2]

In addition to the numerous international laws and international organizations dealing either directly or indirectly with IP, IP has been linked to the right to food. Recognized as a human right under international human rights and humanitarian law, the right to food is established in part by Article 11 of the United Nations' *International Covenant on Economic, Social and Cultural Rights* which imposes an obligation on states to respect existing access to food, protect the right to food, and fulfill the right to food.[3] Jean Ziegler et al. define the right to food as:

> the right to have regular, permanent and unrestricted access, either directly or by means of financial purchases, to quantitatively and qualitatively adequate and sufficient food corresponding to the cultural traditions of the people to which the consumer belongs, and which ensures a physical and mental, individual and collective, fulfilling and dignified life free of fear.[4]

1 Consultative Group on International Agricultural Research, 2012. *Who We Are.* [Online]. Available at: http://www.cgiar.org/who-we-are [accessed: 22 November 2012].

2 Consultative Group on International Agricultural Research, 2012. *CGIAR Principles on the Management of Intellectual Assets*, Article 6.4. [Online]. Available at: http://www.cgiarfund.org/sites/cgiarfund.org/files/Documents/PDF/cgiar_principles_management_intellectual_assets_7march_2012.pdf [accessed: 5 June 2013].

3 By 2011, the right to food was explicitly or implicitly protected in 56 national constitutions, recognized by 51 countries, and acknowledged through framework legislation in a further ten countries: Knuth Lidija, K. and Vidar, M. 2011. *Constitutional and Legal Protection of the right to Food around the World.* Rome: FAO, 32.

4 Ziegler, J., Golay, C., Mahon, C, and Way S.A. 2011. *The Fight For the Right to Food: Lessons Learned.* Geneva: The Graduate Institute.

The United Nations' Special Rapporteur on the right to food, Oliver De Schutter, drew attention to the impact that IP may have on the right to food by publishing in 2009 *The Right to Food: Enhancing Agrobiodiversity and Encouraging Innovation.* In this report, De Schutter examines the impact of seed policies and IP in agriculture on the realization of the right to food. According to De Schutter, when thinking about the interaction between IP and food, international organizations, national governments and policy makers must 'place the needs of the most marginalized groups, including in particular smallholders in developing countries, at the centre of [developing seed policies and IP in agriculture]'.[5] In order to do this, the Special Rapporteur on the right to food made a number of recommendations including that states should: ensure that patent and plant breeders' rights laws do not discourage innovation by introducing barriers to the use of protected material; consider whether access to research tools and plant material is blocked by IP; encourage innovative mechanisms such as patent pools, clearing houses and open source experiments; ensure that seed certification schemes do not exclude farmers' varieties; and support local seed exchange systems such as community seed banks and seed fairs.[6]

While there are numerous texts and commentaries on IP there has actually been very little direct engagement with the interaction between IP and food. To date, scholarship and commentary has focused almost exclusively on access and exchange of genetic resources.[7] By focusing on the 'inputs' to food and crop development (such as plant germplasm) and technological developments (such as genetic engineering), scholarship has tended to focus on *in situ* and *ex situ* collections, as well as the protection of plant varieties, plant genes, genetically modified plants and animals. One reason for the lack of scholarship on food and IP is that there in not a single, coherent area of study known as food and IP. Indeed, IP has multifarious, diverse and subtle interactions with food that makes it difficult, if not impossible, to clearly delineate a field of study, let alone consider all of the possible interactions between IP and food.

Perhaps, most importantly though, giving legal effect to the term 'food' is far from straightforward. Food is central to all aspects of human life. Food has inspired war, motivated the discovery of new worlds, and plays an important role in religion, science, technology and health. Broadly speaking, food includes any substance ingested by humans, is generally of plant or animal origin, and provides nutrients to the human body in the form of carbohydrates, protein, fats, vitamins

---

5 General Assembly. 2009. *Seed Policies and the Right to Food: Enhancing Agrobiodiversity and Encouraging Innovation*, A/64/170. New York: United Nations, 20-21.

6 Ibid., 3.

7 See, for examples, Santilli, J. 2012. *Agrobiodiverstiy and the Law: Regulating Genetic Resources, Food Security and Cultural Diversity*. New York: Earthscan; Lightbourne, M. 2009. *Food Security, Biological Diversity and Intellectual Property Rights*. Farnham: Ashgate.

and minerals. Food is also instrumental in population growth. From the eighteenth to twentieth centuries advances in agricultural production through plant breeding, harvesting and processing meant that human populations thrived. In more recent times the challenge is to balance population increases with food supply. According to the Food and Agricultural Organization of the United Nations (FAO), 975 million people did not have enough food to eat in 2010.[8] Over nine million people die worldwide each year because of hunger and malnutrition, and some three and a half billion people have micronutrient deficiency. The projection that world population, most of which is in developing countries, will grow to an estimated nine billion people by 2050 suggests that the challenge is going to more difficult.

Acknowledging that food is bound up with, and informed by, practices and cultural conventions, Michael Carolan (a sociologist who has written comprehensively on food) suggests that we should be asking 'When is food?', not 'What is food?'[9] This means that when thinking about food and IP, food must be considered more broadly than in absolute, quantitative terms. In his book *Poverty and Famines: An Essay on Entitlement and Deprivation*, Amartya Sen examines a number of famines including the Bengal Famine of 1943, the Ethiopian Famine of 1972–4 and the Famine of Bangladesh in 1974, and argues that it is a loss of capability or entitlement rather than a scarcity of food that leads to famine. At the time, Sen's capabilities and entitlement approach challenged the Malthusian logic that there are simply too many people and not enough food, and in so doing shifted the focus of away from food production towards broader social and political issues that influence peoples' inability to buy food.[10] Indeed, just because we produce more food does not mean that that food gets to the people who need it. This is particularly important when thinking of IP and food, as IP itself is a means of increasing accessibility or reducing accessibility, and not necessarily to actually making *just* more food.

As a result of increasingly complex relations between international and domestic laws, practices and conventions, IP and food interact in infinite ways. These difficulties, however, are not reason enough to overlook the growing importance of IP to food. From our perspective, IP is not just about protecting products or process, brands, information and so on. Nor is it merely about food security and the technological imperative to produce more food. It is a much bigger project. Considering IP and food must be broad enough to include advances in science and technology, human rights, poverty, corruption and the broad expanse of insight from the effects of law and lore in delivering sufficient nutritious 'food' to avoid

8 Food and Agricultural Organization. 2010. *The State of Food Insecurity in the World: Addressing Food Insecurity in Protracted Crises.* [Online]. Available at: http://www.fao.org/docrep/013/i1683e/i1683e.pdf [accessed: 22 November 2012].

9 See for example, Carolan, M. 2013. *Reclaiming Food Security*. New York: Earthscan; Carolan, M. 2011. *The Real Cost of Cheap Food.* New York: Earthscan.

10 Sen, A. 1982. *Poverty and Famines: An Essay on Entitlement and Deprivation.* New York: Oxford University Press.

hunger. In this way the chapters in this book are designed to broadly address IP and food, and cover topics as diverse as geographic indications of tequila, water and rum, 'patent busting', socio-economic factors and the root causes of poverty.

This collection is the output of the IP and Food Workshop held on the Gold Coast, Australia in September 2012 and hosted by Griffith University's Australian Centre for Intellectual Property in Agriculture (ACIPA). The original purpose of the workshop was to collect together academics interested in IP and food issues and to start to traverse the various dimensions of the interaction between IP and food. Each academic was merely asked to address an issue that involves IP and food. The outcome of the workshop reflected in this collection demonstrates that IP has a multifarious and subtle interaction with food, albeit the impacts can be significantly positive and negative. The collection also reflects the immensity of the project that attempts to identify and understand the interactions between IP and food. As such, this collection is one of the early tentative steps along this path.

The collection begins with observations from Professor Robert Henry's laboratory: 'Implications of Advances in Molecular Genetic Technology for Food Security and Ownership'. As Director of Queensland Alliance for Agriculture and Food Innovation and an internationally respected scientist (who has worked in DNA-based methods for the identification of plants, the development of molecular markers for plant breeding and the genetic transformation of plants), Henry is ideally placed to provide the scientists' perspective on IP. Significantly, Henry provides two broad and related insights. First, scientists generally believe that 'rapid advances in DNA sequencing technology and other technology developments have created a revolution in biology that will impact significantly on food production'. Secondly, according to Henry, in order to ensure that science and technology can reach its full potential for food production IP must be 'streamlined' to support emerging scientific and technical opportunities. According to Henry, for example, the use of synthetic DNA potentially necessitates the extension of plant breeders' rights to include protection of DNA sequences to prevent the production of synthetic copies. It also appears, however, that while IP is a concern for scientists, IP itself does not appear to be impeding vital scientific research.

The remaining chapters are then organized in three groups, loosely based on common approaches, arguments or themes. The first group of chapters, 'Public Research and Plant Germplasm: Intellectual Property and Food Security', focus on the increasing role of intellectual property in agricultural research and food production. In 'Reconceptualizing Intellectual Property to Promote Food Security' Brad Sherman examines the place of intellectual property within the CGIAR. Through a careful retracing of intellectual property within the CGIAR system, Sherman shows how there has been a shift towards using intellectual property in publicly funded agricultural research. Sherman claims that the traditional dichotomy between intellectual property and public good research – which has traditionally viewed intellectual property and public good research as mutually exclusive – has been 'unpacked' due to a desire to maximize the impact of agricultural research. The result of which, according to Sherman, has been the reconceptualization of intellectual property as having something to offer in the fight against poverty and food insecurity.

Charles Lawson's 'Intellectual Property Norm Setting in *ex situ* Plant Germplasm Access and Benefit Sharing Arrangements' begins by considering how the CBD and the *Plant Treaty* established global norms for IP in the field of plants, animals and germplasm. Significantly, however, Lawson does not focus on the treaties themselves but shifts attention to five institutions involved in implementing the CBD and the *Plant Treaty*, namely the: Commission on Genetic Resources; the Plant Treaty Governing Body; CGIAR Centres; Global Crop Diversity Trust; and the World Intellectual Property Organization. By contextualizing the practice of access and exchange of plant germplasm, Lawson demonstrates how these institution's policies and materials establish the way in which IP is being used in access and benefit sharing germplasm exchanges. Lawson concludes by claiming that IP may, through establishing norms for the exchange and benefit sharing of germplasm, eventually affect the price and availability of food and feed.

Brendan Tobin's 'Open Access Seeds and Breeds: The Role of the Commons in Protecting Farmers and Livestock Keepers' Rights and Food Security' covers many of the emerging civil society concerns about IP and food and the various dimensions and perspectives that have yet to crystallize into a coherent plan for action. The promise of open access, empowerment of indigenous and local communities, and the recognition of 'rights', promise to deliver just, fair and equitable results.

The next group of chapters, 'Social, Economic and Political Aspects of Food and Intellectual Property', explore some of the factors that are relevant in thinking about how the IP system might operate differently. These chapters principally address matters that are likely to impact on IP, namely social, economic, political and environmental considerations. Berris Charnley's 'Why Didn't an Equivalent to the US Plant Patent Act of 1930 Emerge in Britain? Historicizing the Boundaries of Un-Patentable Innovation' examines the key geographic, political and social differences between the United States (where the *Plant Patent Act* was introduced in 1930) and the United Kingdom (where plants are generally not patentable). By examining the contexts of plant breeding in each country, Charnley argues that it was the mixing together of social factors including plant breeding practices that accounts for the historical difference between the United States and the United Kingdom in providing patent protection to plant breeders. More broadly, Charnley claims that history matters because it hints at why certain innovations may, or may not, be excluded from IP.

Moving from history to regulation, Karinne Ludlow's chapter 'Changing the Recipe: Food Security and Other Socio-Economic Considerations in Agricultural Biotechnology Regulation' takes agricultural biotechnology as her focus. In doing so, Ludlow demonstrates how socio-economic considerations including food security are increasingly included in the decision making processes over the commercialization of genetically modified crops and labeling of genetically modified food. Looking to IP, Ludlow suggests that instruments such as the

Nagoya Protocol place obligations on countries to include socio-economic consideration such as food security in patent regimes. According to Ludlow, adding concerns over food security in the 'recipe' of agricultural biotechnology regulation, including IP, safeguards national sovereignty and consequently plays a role in achieving food security.

The final group of chapters, 'Intellectual Property, Food and Practice', explore specific examples of the interaction between food and IP. In 'Can Intellectual Property Help Feed the World? Intellectual Property, the PLUMPYFIELD® Network and a Sociological Imagination', Jay Sanderson argues that intellectual property currently specifies food security as technology at the expense of the relations in which food is produced, distributed and accessed. Using Nutriset, a France-based company that develops food and nutritional products, as a case study Sanderson illustrates how intellectual property, particularly patents and trademarks, can be used to support locally based food manufacturing and distribution. In this way, Sanderson suggests that intellectual property can act as a vehicle for social change to encourage and support the local production of food and nutritional products which may have long-term benefits for the alleviation of hunger, malnutrition and food insecurity.

Moving from patents and trademarks to geographical indications (GIs), Graham Dutfield's chapter, 'Geographical Indications and Agricultural Community Development: Is the European Model Appropriate for Developing Countries?', asks whether the European approach to GIs could be used by developing countries. Already having a limited presence in countries such as Brazil, Mexico, Cuba, Algeria and Tunisia, GIs – which can be used as a marketing and branding tool that allows consumers to identify goods as origination in a specific territory, region or locality – may empower local communities, and stimulate employment and training. Taking the small Caribbean Island of Saint Lucia as his focus, Dutfield explores how GIs may play a role in increasing the incomes of local communities and small-scale producers (rum and beer), improving employment (handicrafts) and contribute to the diversification of local agricultural production (cocoa and cocoa products).

Matthew Rimmer's 'Patent-Busting: The Public Patent Foundation, Gene Patents and the Seed Wars' considers whether the Public Patent Foundation, a United States-based, non-profit organization that attempts to limit abuse of United States patent law by 'representing the public's interests in the patent system', could be a model for challenging problematic patents. For example, Rimmer examines the so-called 'seed wars' in which the Public Patent Foundation challenged, initially in the United States District Court, Monsanto's patents in respect of genetically modified organisms. Interestingly, Rimmer points out that although 'patent busting' is generally limited to the United States and to subject matter such as human genes, medical information and information technology, it has the potential to play a role in challenging not just food-related patents, but also food-related plant breeders' rights and trademarks.

Taken together, these various contributions show that IP and food are intimately associated and that IP has a distinct and important role in both promoting and restricting the production, availability and access to food. The contributions also demonstrate that the IP and food project is much bigger. The broader project must also include advances in science and technology, human rights, poverty, corruption and the broad expanse of insight from the effects of law and lore in delivering sufficient nutritious 'food' to avoid hunger. Indeed, the IP and Food project is only beginning.

PART I
Observations from the Laboratory

Chapter 2
Implications of Advances in Molecular Genetic Technology for Food Security and Ownership

Robert J. Henry

Introduction

Recent rapid advances in DNA sequencing technology and other technology developments have created a revolution in biology that will impact significantly on food production and the underlying crop and animal breeding. These developments will facilitate accelerated exchange of genetic materials and data but will require a streamlined IP environment if we are to realize the improved food security and biodiversity conservation that this technology promises.

Technology Developments

With continued strong growth in demand for food due to population growth and growth in per capita consumption (due to economic development), food security relies on continued and sustained genetic improvement of the organisms (plants, animals and microorganisms) that form the basis of food production.[1] We need to double food production from 2000 production levels by 2050 to cope with the around 50 per cent expected growth in population and the greatly increased consumption per person. Significantly, this increase in food production will need to be delivered from the same, or a reduced area of agricultural land, which means that food production will need to double per unit area of land. The threat of climate change puts additional demands on the food system.

Biodiversity conservation also remains crucial to food security. Food production depends on biodiversity as a source of new resources to breed agricultural varieties with the productivity characteristic required. This has historically supplied much of the growth in agricultural productivity. Biodiversity conservation also provides security for the ecosystem services necessary to sustain agriculture and food production.

1 Henry, R.J. (2010) *Plant Resources for Food, Fuel and Conservation*. London: Earthscan.

Technical developments in genetic technology such as advances in DNA sequencing, chemical synthesis of DNA and DNA (gene) transfer between organisms, are providing new tools that could be used to address some of these challenges. The IP system needs to adapt to support the technical opportunities that are emerging with the potential to deliver more sustainable food production that could contribute significantly to meeting the global challenge of food security. In the following pages I provide a scientists' perspective on IP, and outline some of the key scientific and technological advances relevant for food production; pointing out that, from a scientists' perspective at least, science and technology raise a number of important issues for IP protection, infringement and enforcement. More specifically, I provide some thoughts on IP and technological development, changes in taxonomy and the implications of science and technology on food security, biodiversity conservation and food labelling. Lastly, I offer some thoughts on IP regimes and laws relevant to access and exchange of genetic resources. For those of us in the laboratory: IP must be 'streamlined' to support emerging scientific and technical opportunities.

DNA Sequencing

Rapid and continuing advances in DNA sequencing technology[2] have created an explosion of biological information in the last few years.[3] These technical advances have exploited advances in nanotechnology and information technology to deliver automated platforms for large-scale DNA sequence data analysis. With the capture of all of the DNA sequence information in an organism it is increasingly possible, and cost effective, as a tool for increasing biological understanding, genetic improvement for food production and even in the enforcement of IP. The new sequencing technologies have been applied in different ways to achieve specific outcomes.[4] Whole genome sequencing, for instance, allows all the genetic data of a species or variety to be collected,[5] which means that genetic improvement for food production and enforcement of IP is more cost-effective for large numbers of samples. Furthermore, sequencing the gene-rich parts of the genome[6] (rather than the entire species or variety) can be achieved by genome enrichment before

2 Edwards, D., Henry, R.J. and Edwards, K.J. 2012. Advances in DNA sequencing accelerating plant biotechnology. *Plant Biotechnology Journal*, 10(6), 621-622.

3 Henry, R.J. 2013. Evolution of DNA marker technology in plants, in *Molecular Markers in Plants*, edited by R.J. Henry. Ames and Chichester: Wiley-Blackwell, 3-19.

4 Edwards, M.A. and Henry, R.J. 2011. DNA sequencing methods contributing to new directions in cereal research. *Journal of Cereal Science*, 54(3), 395-400.

5 Gopala Krishnan, S., Waters, D.L.E., Katiyar, S.K., Sadananda, A.R., Satyadev, V. and Henry, R.J. 2012. Genome-wide DNA polymorphisms in elite indica rice inbreds discovered by whole-genome sequencing. *Plant Biotechnology Journal*, 10(6), 623-634.

6 Bundock, P.C., Casu, R. and Henry, R.J. 2012. Enrichment of genomic DNA for polymorphism detection in a non-model highly polyploidy crop plant. *Plant Biotechnology Journal*, 10(6), 657-667.

sequencing complex genomes.[7] The functional parts of the genome can be analysed specifically by sequencing the expressed genes.[8] Very specific sequencing of genes in particular biochemical pathways[9] or areas of metabolism[10] can be achieved by amplicon sequencing in very large populations, if necessary, to associate gene sequence with traits. Indeed, the technology is rapidly making conventional gene cloning redundant. The technology is generating protocols for routine and definitive identification of biological samples[11] that are much more informative than earlier identification tools based upon the analysis of individual genes.

DNA sequencing raises a number of issues for IP. The protection of genetic IP by genotyping can now provide the ultimate protection of the entire genome sequence. The policing of IP by genetic testing are greatly advanced by these developments. Wild genetic resources provide an important source of diversity for use in crop improvement.[12] These new tools allow much better characterization of genetic resources for food production.[13] Analysis of diversity in wild populations growing across environmental gradients has been used to examine how wild populations are adapting to climate differences,[14] providing important clues to strategies for adaptation of agriculture to climate change.

---

7 Souza, G.M., Berges, H., Bocs, S., Casu, R., D'Hont, A., Ferreira, J.E., Henry, R., Ming, R., Potier, B., Van Sluys, M.A., Vincentz, M. and Paterson, A.H. 2011. The sugarcane genome challenges: Strategies for sequencing a highly complex genome. *Tropical Plant Biology*, 4(3), 145-156.

8 Gillies, S., Furtado, A., Henry, R.J. 2012. Gene expression in the developing aleurone and starchy endosperm of wheat. *Plant Biotechnology Journal*, 10(6), 668-679.

9 Kharabian-Masouleh, A., Waters, D.L.E., Reinke, R.F., Ward, R., Henry, R.J. 2012. SNP in starch biosynthesis genes associated with nutritional and functional properties of rice. *Scientific Reports*, 2, 557; Kharabian-Masouleh, A., Waters, D.L.E., Reinke, R.F. and Henry, R.J. 2011. Discovery of polymorphisms in starch related genes in rice germplasm by amplification of pooled DNA and deeply parallel sequencing. *Plant Biotechnology Journal*, 9(9), 1074-1085.

10 Sexton, T.R., Henry, R.J., Harwood, C.E., Thomas, D.S., McManus, L.J., Raymond, C., Henson, M. and Shepherd, M. 2012. Pectin methylesterase genes influence solid wood properties of *Eucalyptus pilularis*. *Plant Physiology*, 158(1), 531-541.

11 Nock, C., Waters, D.L.E., Edwards, M.A., Bowen, S., Rice, N., Cordeiro, G.M. and Henry, R.J. 2011. Chloroplast genome sequence from total DNA for plant identification. *Plant Biotechnology Journal*, 9(3), 328-333.

12 Lakew, B., Henry, R.J., Ceccarelli, S., Grando, S., Eglinton, J. and Baum, M. 2012. Genetic analysis and phenotypic associations for drought tolerance in Hordum spontaneum introgression lines using SSR and SNP markers. *Euphytica*, in press, accepted 13 March 2012.

13 Waters, D.L.E., Nock, C.J., Ishikawa, R., Rice, N. and Henry, R.J. 2012. Chloroplast genome sequence confirms distinctness of Australian and Asian wild rice. *Ecology and Evolution*, 2(1), 211-217.

14 Shapter F.M., Fitzgerald, T.L., Waters, D.L.E., McDonald, S., Chivers, I.H. and Henry, R.J. 2012. Analysis of adaptive ribosomal gene diversity in wild plant populations from contrasting climatic environments. *Plant Signaling & Behavior*, 7(1), 1-3; Fitzgerald,

Synthetic DNA

The ability to chemically synthesis large DNA molecules creates the potential for complete synthesis of living organisms without the use of biological material. Commercial services are now available to produce a DNA molecule with the sequence you supply. This has significant implications for IP issues associated with the plants, animals and microbes used for food production. Perhaps most significantly, the protection of plant varieties may need to be extended to include protection of their DNA sequence data to prevent the production of synthetic copies.

DNA Transfer between Organisms

The technology to move genetic material between organisms continues to improve. Direct and agrobacterium mediated transformation have both been improved for many species. DNA from another organism or completely synthetic DNA based upon data from a biological system or of novel design can be introduced to genetically modify the target organism. IP laws need to recognize this technical advance and the implications for protection of varieties and genes.

Domestication of New Organisms

New sequencing technologies can be applied to identification of genes that need to be selected to achieve domestication for use in agriculture and food production.[15] This technology may facilitate the rapid domestication of new species to satisfy the demands of food security.[16] New cereal crops better adapted to some production environments would be an example of the potential application of this technology. In terms of IP, new crop species may have different IP constraints than traditional crop species. Many of these species will be outside the scope of IP agreements designed to cover existing domesticated species.

T.L., Shapter, F.M., McDonald, S., Waters, D.L.E., Chivers, I.H., Drenth, A., Nevo, E. and Henry R.J. 2011. Genome Diversity in Wild Grasses under Environmental Stress. *Proceeding of the National Academy of Science USA*, 108: 21139- 21144.

15 Malory, S., Shapter, F.M., Elphinstone, M.S., Chivers, I.H. and Henry, R.J. 2011. Characterizing homologues of crop domestication genes in poorly described wild relatives by high-throughput sequencing of whole genomes. *Plant Biotechnology Journal*, 9(9), 1131-1140.

16 Henry, R.J. 2012. Next generation sequencing for understanding and accelerating crop domestication. *Briefings in Functional Genomics*, 11(1), 51-56.

Impact of Changes in Taxonomy

Biologists classify organisms using the system of binomial nomenclature. However taxonomists often reclassify organisms based upon new discoveries or further research. DNA analysis is now a key tool used in supporting these changes. IP agreements or protection of genetic resources needs to keep pace with these developments. The names of the species covered in agreements may change and this may lead to confusion and sometimes a failure to protect IP. For example, the genus name changing could result in the use of a name that was not on the list of species covered by the FAO *Plant Treaty* or on the lists of endangered species.

Implications for Food Security

Food security relies on the development of technologies to allow food production to keep pace with demand. As noted above, two factors are driving strong growth in food demand globally. First, continued population growth creates more mouths to feed. Secondly, and equally important, is economic development, especially in Asia, resulting in growth in food consumption per capita as living standard rise resulting in a greater ability to afford food.

Energy security is a related issue. Energy is required for both food production and distribution. As fossil fuel reserves, especially oil, are exhausted and the environmental consequences of rising returning the carbon in these reserves to the atmosphere are better understood, the need for alternative sources becomes more evident. Bioenergy has the potential to replace fossil fuels and make a contribution to energy security and environmental protection. This requires that new technologies for bioenergy production are developed and delivered in a way that avoids competition with food production and contributes positively to food production by reducing the cost and environmental impact of energy inputs to food production. New species may be needed to deliver sustainable bioenergy production.[17]

Genetic technologies have been a major contributor to meeting food requirements in the past and are likely to remain a major contributor to food security in the future. We have two options for improving food production, we can improve the genetics of the food organism or we can better manage the environment of the food organism to deliver more food. We need to do both successfully to be able to achieve adequate food security. Evidence from the past suggests that these two factors differ in importance in relation to specific food commodities and time periods. Overall, it is likely that they have been about equally important and this is the best guess as to the future requirements.

---

17 Shepherd, M., Bartle, J., Lee, D.J., Brawner, J., Bush, D., Turnbull, P., Macdonell, P., Brown, T.R., Simmons, B. and Henry, R.J. 2011. Eucalypts as a biofuel feedstock. *Biofuels*, 2(6), 639-657.

The new sequencing technologies have delivered a major breakthrough in the discovery of genetic markers for use in genetic improvement of agricultural species facilitating improved food security.[18] This technology should contribute to a significant improvement in the effectiveness of plant breeding for major and minor crops and for animal species.

Implications for Biodiversity Conservation

Biodiversity conservation competes with agriculture for space on land. Agriculture occupies a very high proportion of the arable land on earth. It is important that we deliver food security by improving the efficiency of agriculture rather than increasing the footprint (area of land used) of agriculture because of the implications for biodiversity conservation. Highly efficient agriculture will minimize the land footprint and allow space for biodiversity conservation. The importance of biodiversity conservation for food production has been explained above, providing a source of diversity for genetic improvement of agricultural species and providing 'ecosystem services' supporting the environment required for life on earth. Biodiversity is also important for other more abstract reasons. It makes life more interesting (if humans were the only life form some would argue that life might not be so interesting).

New genetic technologies can make a significant contribution to biodiversity conservation. DNA analysis tools help better define levels of biodiversity and improve our understanding of evolutionary processes. Management of genetic resources, both *in situ* and *ex situ*, can be informed by this knowledge. This helps both biodiversity conservation and the use of biodiversity to support agriculture and food production. The management of IP associated with biodiversity will be important for both of these related objectives.

Implications for Food Labelling

DNA testing can be used to identify the species or varieties of organisms in the food. As this capability becomes more affordable and widespread policing of IP rights associated with foods may become more active and effective. The technology can be used to protect plant patents, gene patents, plant breeders' rights and the use of geographical indicators. Often food producers will want to use this approach to characterize new processed food products of competitors to police their IP rights and to support their own product development activities.

---

18 Henry, R.J., Edwards, M., Waters, D.L.E., Krishnan, S.G., Bundock, P., Sexton, T.R., Masouleh, A.K., Nock, C.J., Pattemore, J. 2012. Application of large scale sequencing to marker discovery in plants. *Journal of Bioscience*, 37(1), 1-13.

IP Barriers and Incentives to Applying this Technology for Enhanced Food Security and Biodiversity

Food security relies on an IP system that encourages innovation while protecting the resource base (biodiversity) for food production. Technology developments may require new approaches to IP management to ensure these outcomes.

Patents

Gene patents have been important for transgenic organisms and have served the interests of promoting this area well. Many patents on enabling technology such as gene transfer have been in force for some time and will end soon. Patents on specific genes or gene regulatory elements will be important for some time. However, many transgenic crop varieties are being developed with many 'stacked' genes providing many useful traits in the one transgenic organism. This requires freedom to operate with many genes and gene promoters for each product. The value of individual gene patents is likely to be lower in this market environment and non-exclusive licensing may become more widespread.

PBR

Plant Breeders' Rights (PBR; consistent with UPOV) has provided a useful IP protection for plants with commercial value in many countries. The level of protection varies and further harmonization and expansion of this system would be helpful. The presence of transgenes protected by patents within a plant variety protected by PBR will complicate IP issues. The requirement that PBR-protected material is made available for further improvement may prove difficult when the variety contains patented technology in the form of transgenes. Breeders using this material may need to licence transgenes contained within the material to be able to commercialize derived varieties.

MTAs for Genetic Material

The control of access to genetic resources remains challenging, with ongoing attempts to encourage use of a standard MTA (Material Transfer Agreement). The differ status of species covered by the *Convention on Biodiversity* and the *Plant Treaty* (see below) complicates attempts to have a standard MTA for all materials.

Seeds

Very large international and national seed collections are held for major food crops. Exchange of seed between these collections and between collections and researchers and plant breeders is controlled by MTAs. The explosion in the amount of sequence data will impact on these collections by identifying duplicates (especially those held in separate collections or the same collection under different names) and devaluing some collections and demonstrating the uniqueness of other material and increasing the perceived value of this material. Rationalization of *ex situ* genetic resource collections should be facilitated by these developments.

Genes

MTAs are usually used for exchange of DNA samples between laboratories. These are usually exchanged as a clone in a vector allowing propagation in bacteria. Improvements in DNA synthesis technology makes it possible to exchange data on the sequence and allow synthesis avoiding transfer of a physical sample. MTAs usually relate to the physical sample. The DNA sequence is often disclosed or not able to be protected. IP associated with clones is probably of declining value because of these technology changes and the practice of using MTAs for these transfers may decline.

MTAs for Data

The exchange of some data sets is also controlled by MTAs. These data sets are sometimes restricted to members of a research consortium. This approach is less likely in the future as much more data comes into the public domain and the cost of data collection declines. Many groups of genome scientists have formed consortia to work together on the analysis of important species. The International Coffee Genome Consortium is a typical example of such a group. Scientists from many countries (including Australia, Brazil, Canada, France, India, Italy and the USA) have agreed to exchange data to allow a reference genome sequence to be created for coffee. In this case data is shared between the participants until a reference sequence is produced and analysed, resulting a scientific publication that puts all the data in the public domain.

International Agreements on Exchange of Genetic Resources

Several international agreements influence the exchange of genetic resources:

- *Convention on Biodiversity* – The biodiversity convention empowers countries to claim ownership of their genetic resources.

- *FAO Plant Treaty* – The FAO Treaty ensures free access to genetic resources for major food species. Species in this treaty are defined as genera giving the treaty wide coverage.
- *The Nagoya Biodiversity Agreement* – The recent Nagoya Protocol (2010) seeks collaboration in the protection of biodiversity *in situ*.

The implementation of these international agreements will require ongoing education of those involved in the collection, storage management and utilization of genetic resources in all countries. The development of efficient and effective management systems to achieve the goals of these agreements is important if food security and resource conservation is to be delivered.

DNA Banks

A network of DNA banks[19] has been created to facilitate the analysis and utilization of biodiversity. DNA banks play an important role in allowing exchange of samples of DNA for analysis without the need to move seeds or living materials.[20] This avoids quarantine issues and reduces demand on limited seed stocks as many groups can analyse the same DNA sample from a seed in a collection. A plant breeder can request some DNA from the DNA bank and screen this for the presence of the desirable genes before requesting seed from the seed collection. Large numbers of samples can be screened at the DNA level and seed requested only for those of interest. The IP issues associated with DNA are not always clear. Do the IP rights associated with the organism extend to the DNA derived from that organism or can DNA samples be exchanged without reference to the IP arrangements controlling exchange of the organism itself? A network that provides linkages and standards for data collection and exchange been established for DNA banks. Coordination of DNA banking in this way should assist in joint consideration of IP issues.

Importance of Global Networking

Effective exchange of materials and data in support of food security relies on contact between organizations and individuals. International networking or researchers is a key to effective sharing of enabling IP. The CGIAR (Consultative Group on International Agricultural Research) is an important framework for

19 DNA Bank Network, *DNA Bank Network – Home*. [Online]. Available at: www.dnabank-network.org/ [accessed: 3 January 2013].

20 Rice, N., Henry, R.J. and Rossetto, M. 2006. *Plant DNA Banks – A Primary Resource for Conservation Research*, in *DNA Banks – Providing Novel Options for Genebanks?* edited by M.C. de Vicente. Rome: International Plant Genetic Resources Institute, 41-48.

international collaboration. Major international conferences such as the Plant and Animal Genome Conference provide important forms for the development of international collaboration.

ICRCGC

An International Climate Resilient Crops Genomics Consortium (ICRCGC)[21] has been formed to promote application of genomics and sharing of technologies. The potential of genomic technologies to deliver their potential outcomes as required for global food security seem unlikely without significant international coordination and collaboration.

Public Intellectual Property Resource for Agriculture

The Public Intellectual Property Resource for Agriculture (PIPRA) was established by the Rockefeller Foundation in 2004 and aims to supports access to critical IP for public sector research and development to advance crop improvement globally.[22] PIPRA provides access to freedom to operate with plant transformation technology. Similar arrangements might be useful for other IP in this area.

Need for the IP System to Adapt to Technology Advances

Our ability to easily obtain DNA from an organism and to then determine the sequences of the DNA continues to improve. DNA sequence data can be easily transferred (using the internet) or shared and linked to technologies that allow DNA synthesis and incorporation of the DNA into another organism. IP law needs to recognize the relationship between living organisms, their DNA and the data defining the DNA sequence. The interchangeability of this data and materials needs to be considered in drafting effective IP arrangements.

---

21 International Climate Resilient Crops Genomics Consortium, *Climate Change Genomics*. [Online]. Available at: www.climatechangegenomics.org [accessed: 3 January 2013].

22 Chi-Ham, C.L., Boettiger, S., Figuerosa-Balderas, R., Bird, S., Geoola, J.N., Zamora, P., Alandete-Saez, M. and Bennett, A.B. 2012. An intellectual property sharing initiative in agricultural biotechnology: development of broadly accessible technologies for plant transformation. *Plant Biotechnology Journal*, 10(5), 501-510.

PART II
Public Research and Plant Germplasm: Intellectual Property and Food Security

Chapter 3
Reconceptualizing Intellectual Property to Promote Food Security

Brad Sherman

Introduction

Typically, intellectual property is seen as a resource that is used by private sector (commercial) organizations to recoup investment in innovation, as a mechanism to exploit creative outputs, or as a means to protect business reputation and goodwill. From this perspective, there is little scope for intellectual property outside of the commercial sphere; at best it is able to protect the reputation of authors and the integrity of their works, at worst it positively hinders public good goals. While this logic has been applied in many contexts, it has been particularly prevalent in relation to publicly funded agricultural research that aims to improve food security and reduce poverty. Under this stereotypical view, there is little that intellectual property can do to assist these goals: at best it is irrelevant, at worst it is a positive hindrance. Despite the prevalence of this way of thinking, there has been a gradual transformation in the way that intellectual property law operates in relation to publicly funded agricultural research. While a number of changes have occurred in developed countries, the changes have been more marked in relation to research that focuses on food security in developing countries. While the goals of poverty reduction and food security remain the same, the attitude towards intellectual property has changed. The aim of this chapter is to chart the nature of this transformation. In so doing it looks at the way that the (apparently) conflicting goals of intellectual property law, which are typically associated with stimulating private sector investment in research, and the goals of improving food security and reducing poverty which underpin much publicly funded development-focused agricultural research have been reconciled. More specifically, it will show that in resolving this tension there has been a change in the way that research and intellectual property are conceptualized.

To do this, this chapter will focus on the fate of intellectual property within CGIAR (Consultative Group on International Agricultural Research). CGIAR was established in 1971 as a strategic alliance of four international agricultural research centers and a limited number of donors who had a shared vision to produce research that improved food security and reduced poverty and hunger.[1] This vision was underpinned by the idea that all of the research results generated by Centers should be freely available without restriction and as widely as possible. While the overarching goals have largely remained the same, over the last 50 years the number of Centers has expanded to 15, membership has increased from 18 (countries and organizations) to over a hundred, and the number of fund donors has expanded considerably. Centers are now members of the CGIAR Consortium, which has been established as a legal entity in its own right.

Intellectual Property in CGIAR

When thinking about the role that intellectual property plays within CGIAR, it is important to keep in mind that researchers and Centers interact with intellectual property in a number of different circumstances. It is also important to bear in mind that there is no unified approach to intellectual property. While it may be possible to group Centers together – for example, those with considerable experience and expertise – there are differences in the way that Centers approach intellectual property. In some cases, there are also differences within Centers as to how they think about and deal with intellectual property. While these differences mean that we must be careful about the generalizations we make about intellectual property and the role it plays within CGIAR, it is still possible to divide the approach taken towards intellectual property in CGIAR into three stages.

In the early period (1971–80), intellectual property was largely alien to the day-to-day activities of the Centers and their staff. Centers rarely (if ever) sought or condoned intellectual property protection over the results of their research. Neither CGIAR officials nor Center members regularly engaged in policy debates about intellectual property, nor did they publicly discuss the pros and cons of, for example, patent protection for biological resources.[2] While there were some exceptions, for the most part intellectual property was something that simply did not that figure in any significant way in CGIAR.

The situation began to change in the early 1980s, however, as intellectual property slowly became something that CGIAR had to grapple with. Gradually,

---

1 For a general history see Özgediz, S. 2012. *The CGIAR at 40: Institutional Evolution of the World's Premier Agricultural Research Network*. Washington: CGIAR; Ryan, J.G. 1990. *The Evolution of TAC's Position on Intellectual Property Rights and the IARCS: Some Notes from the Record*. [Online]. Available at: http://www.agric-econ.uni-kiel.de/Abteilungen/II/forschung/file3.pdf [accessed: 7 February 2013]; Pistorius, R. 1997. *Scientists, Plants and Politics: A History of the Plant Genetic Resources Movement*. Rome: International Plant Genetic Resources Institute.

2 One of the earliest examples occurred at a March 1980 meeting with European CGIAR donors in Switzerland where Jap Hardon raised problems about access to genetic resources and recommended a study of plant breeders' rights for developing countries.

intellectual property began to appear on the agendas of CGIAR committees, to be debated in meetings of Center Directors, and discussed in CGIAR reports and public communiqués. For most of the 1980s the primary concern was plant breeders' rights and the possible stifling effect that these rights may have had on access to plant genetic resources.[3] At this time, the approach to intellectual property was clear and straightforward: intellectual property had no real role to play either in the Centers or the CGIAR system more generally. As publicly funded institutions, the results of the research and development activities of the Centers were seen as international public goods that were, and ought to remain, freely available for all to use. Within this framework, there was little or no room for intellectual property, which was seen as the exclusive tool of private sector commercially focused organizations: at best intellectual property did little to assist the goals of poverty reduction; at worst, it hindered and restricted access, pushed up prices and undermined the goals of CGIAR.

Over time, the traditional approach towards intellectual property has slowly been undermined.[4] As the external review of the Centro Internacional de Agricultura Tropical (CIAT) said in relation to intellectual property management in the Center, '[w]ithin a short span of time, the "rules of the game" have changed'.[5] While intellectual property is not and will not become a central factor in the way that research in CGIAR is managed, intellectual property has taken on a much more prominent and proactive role in a number of areas.[6] Specifically, intellectual property has become more important in terms of the extent to which Centers make

3 See, for example, Center Directors and Center Board Chairs. 1999. *CGIAR Center Statements on Genetic Resources, Intellectual Property Rights, and Biotechnology*. [Online]. Available at: http://library.cgiar.org/bitstream/handle/10947/253/mtm9805e.pdf?sequence=1 [accessed: 7 February 2013].

4 It is possible to see the change of focus in the shift from earlier reports – such as the 1982 report by Aberg, E. 1982. *The Relationship Between Plant Breeding, Germplasm Collection and Plant Breeders' Rights*, IBPGR 81/16. Rome: IBPGR Secretariat; and the 1982 Technical Advisory Committee. 1982. *TAC Statement on Plant Breeders' Rights and International Agricultural Research Centers*, AGD/TAC:IAR/82/23. Rome: TAC Secretariat – to later reports such as 1996 *CGIAR Center Statements*, above n 3; and the 1996 *Guiding Principles for the Consultative Group on International Agricultural Research Centers on Intellectual Property and Genetic Resources* in System-wide Genetic Resources Programme. 2003. *Booklet of CGIAR Centre Policy Instruments, Guidelines and Statements on Genetic Resources, Biotechnology and Intellectual Property Rights*, Version II. Rome: System-wide Genetic Resources Programme, 30-32 (republished and updated in 2010).

5 TAC Secretariat. 2001. *Report of the Fifth External Programme and Management Review of the Centro Internacional de Agricultura Tropical (CIAT)*. Rome: TAC Secretariat, 122.

6 One of the comments sometimes made about the Technical Advisory Committee (TAC), which was the focal point for discussions about intellectual property within the CGIAR in the 1980s, was that it retained a passive, watching brief on intellectual property.

use of products and processes protected by intellectual property, in relation to the management of genetic resources controlled by the Centers, and as a means to protect the research undertaken within CGIAR. I will look at each in turn.

The first area where there has been a shift away from the traditional approach to intellectual property is the extent to which Centers make use of and rely upon things that are protected by third party intellectual property (often described as a growing use of 'proprietary inputs').[7] The growing use of objects protected by third party intellectual property covers a range of products and processes from patent protection over research tools (such as polymerase chain reaction (PCR)), diagnostic test kits, genes, and new varieties of plants through to copyright protection over computer programs, data bases, manuals, journal articles and the design of greenhouses. Strictly speaking, this is not new: Centers have always made use of objects protected by intellectual property. What has changed, however, is the extent of use. In many instances this was prompted by the extension of intellectual property protection over upstream research (particularly in relation to biotechnological inventions), by the use of more expensive and sophisticated laboratory equipment developed and protected by the private sector, and by subtle changes prompted by intellectual property (such as the move from ownership to licensing of computers and computer software to get around problems of exhaustion). While these new uses have, at times, created some disquiet, they have quickly been subsumed into the everyday activities of the Centers.

The second area where there has been a change in the way that the Centers and the CGIAR system more generally interact with intellectual property is in relation to the genetic resources that are under the stewardship of the Centers. The various collections held by the Centers have underpinned much of the research undertaken by CGIAR researchers, particularly the development of new varieties. At a general level, the approach taken to these genetic resources remains the same: the Centers operate on the basis that they hold the genetic resources in trust for the global community. What has changed, however, is the way that this overarching aim has been formulated and how it is executed. While the nature of these changes has been more subtle and indirect than some of the other changes that have taken place, nonetheless they still represent an important transformation in the role that intellectual property (and related legal rights) play within CGIAR.

A number of factors have coalesced to bring about these changes. One important event occurred in 2005 when two Australian organizations attempted to take out plant breeders' rights protection over two varieties of chickpea that

7 See International Service for National Agricultural Research. 1998. *Proprietary Biotechnology Research Inputs and international Agricultural Research: Report of ISNAR Study Commissioned by CGIAR Expert Panel on Proprietary Science*, Briefing Paper 39. The Hague: International Service for National Agricultural Research. See also Cohen, J., Falconi C. and Komen J. 1999. Perspectives from International Agricultural Research Centers, in *Intellectual Property Rights in Agriculture: The World Bank's Role in Assisting Borrower and Member Countries,* edited by U. Lele et al. Washington: World Bank, 23-25.

had been supplied to them by the CGIAR's International Center for Agricultural Research in Dry Areas (ICARDA). While the applications were ultimately withdrawn, nonetheless they still played an important role in heightening public awareness of some of the potential problems associated with intellectual property protection.[8] The attempt to protect unimproved material that was meant to have been held in trust by ICARDA also had long-term consequences for the Center's ability to collect genetic materials.

Another factor that changed the role that intellectual property plays in relation to genetic resources within CGIAR was the *International Treaty on Plant Genetic Resources for Food and Agriculture* (*Plant Treaty*), which entered into force in 2004. While the Treaty only applies to a proportion of the genetic resources held by the Centers, nonetheless it has been adopted as the model by which *all* of the genetic resources held in Center collections are managed.[9] The Treaty introduced three general changes in the way that Centers deal with genetic resources. The first was in relation to the principle of open access. While the content of this ideal has not changed, at least not within CGIAR, the form that it takes has. Previously, although this well-known principle was widely adhered to, it was not formally recognized within CGIAR. The situation changed, however, as a result of the 'in-trust' agreements that the Centers signed with the FAO in 1994, which constituted formal legal recognition of the ideal of open access.[10] The agreements, which were seen as a stop-gap measure until the treaty was concluded, were designed to provide certainty about the approach of Centers towards genetic resources. They also brought the collections into line with the 1992 *Convention on Biological Diversity*. To this end, the agreements provided that Centers held 'designated

8 See ETC Group. 1988. *Recent Australian Claims to Indian and Iranian Chickpeas Countered by NGOs and ICRISAT*. [Online]. Available at: http://www.etcgroup.org/es/content/recent-australian-claims-indian-and-iranian-chickpeas-countered-ngos-and-icrisat [accessed: 7 February 2013]; ETC Group. 1988. *International Research Center (ICARDA) Breaks Trust*. [Online]. Available at: http://www.etcgroup.org/content/international-research-centre-icarda-breaks-trust [accessed: 7 February 2013]. For the CGIAR response see CGIAR Press Release. 1998. *CGIAR Urges Halt to Granting of Intellectual Property Rights for Designated Plant Germplasm*. [Online]. Available at: http://library.cgiar.org/bitstream/handle/10947/357/n98002.pdf?sequence=1 [accessed: 7 February 2013].

9 A preliminary question concerned the legal status of the CGIAR germplasm collections. On this see Siebeck, W. and Barton, J. 1991. *The Legal Status of CGIAR Germplasm Collections and Related Issues*. [Online]. Available at: http://library.cgiar.org/bitstream/handle/10947/1549/tc9203b.pdf?sequence=1 [accessed: 7 February 2013].

10 See Consultative Group on International Agricultural Research. 1994. *Joint Statement of FAO and the CGIAR Centers on the Agreement Placing CGIAR Germplasm Collections under the Auspices of FAO*. [Online]. Available at: http://library.cgiar.org/bitstream/handle/10947/697/cg9410s.pdf?sequence=1 [accessed: 7 February 2013]; CGIAR Press Release, 1988. *World Plant Collections Placed under Intergovernmental Auspices*, Press ICW/94/3. [Online]. Available at: http://www.cgiar.org/web-archives/www-cgiar-org-newsroom-releases-news-asp-idnews-200/ [accessed: 7 February 2013].

germplasm in trust for the benefit of the international community'. The 1994 agreements also provided that Centers were under an obligation to 'make samples of the designated germplasm and related information available directly to users or through FAO, for the purpose of scientific research, plant breeding or genetic resources conservation, without restriction'. The second change instigated by the *Plant Treaty* was in terms of the way that Centers manage their genetic resources internally. This occurred because the in-trust agreements that pre-empted the Treaty stipulated that the Centers would not seek intellectual property protection over the genetic material materials that they held in their various collections. While it was highly unlikely that this was going to occur, nonetheless the declaration provided an important statement of intent about the role and place of intellectual property within the CGIAR. The third change brought about by the Treaty related to the way that Centers manage and control what happens to their genetic collections when they are transferred to third parties. This occurred as a result of the 2007 decision by the Centers to adopt the Standard Material Transfer Agreement (SMTA) that had been developed by the Governing Body of the Treaty to manage the transfers of genetic materials covered by the Treaty (namely, the genetic materials listed in Annex 1 of the Treaty). Importantly, this agreement extended beyond the 64 materials listed in Annex 1 to include non-Annex 1 materials collected before the entry into force of the Treaty.[11] There are two noteworthy features of the SMTAs that are used when Centers transfer genetic materials to third parties; firstly, they specify that intellectual property protection cannot be taken out on materials 'in the form received' by a third party recipient;[12] and, secondly, in the event that a recipient improves and commercializes material received from a Center *and* the improved material is not available to others for research or breeding, the material transfer agreements provide that the recipient must make the material available to other parties in accordance with the principles of fair and equitable benefit sharing. The material transfer agreements also require the recipient to pay a percentage of any income derived from the improved genetic resource to a central fund that is used to maintain the germplasm collections.

The third area where there has been a change in the role that intellectual property plays within CGIAR is in respect to the research results – the data,

---

11 In 2007, the Centers agreed to distribute 'in trust' Annex 1 materials (the 64 crops and foraged listed in Annex 1 to the Treaty) using the standard material transfer agreement (SMTA) that had been adopted by the Governing Body of the Treaty as the basis on which all transfers of materials in the Treaty's multilateral system of 'access and benefit sharing'. The Centers also agreed to use a material transfer agreement amended by the Governing Body during its second session when distributing non-annex 1 materials. This aberration was corrected in 2007 when it was decided that the Centers should use the same SMTA when distributing non-annex 1 materials. System-wide Genetic Resources Programme (SGRP). 2009. *Guide for the CGIAR Center's Use of the Standard Material Transfer Agreement.* Rome: Bioversity International, 2.

12 *Standard Material Transfer Agreement*, Article 6.2.

products and processes – generated by researchers within the CGIAR system. Historically, Centers and their staff have, by default, been owners of the copyright in the various works that they create: notably, in relation to research reports, journal articles and books. This is not the case, however, with respect to those types of intellectual property that are dependent on registration – notably patents, plant breeder's rights, and design protection – where there has been very little active engagement. While it is difficult to work out how much CGIAR research has been protected by intellectual property, it is clear that intellectual property protection for CGIAR research – whether by researchers, Centers, or by third parties in consultation and agreement with Centers – has increased. There are a number of reasons for this transformation.

One factor that played a role in increasing CGIAR involvement with intellectual property is the fact that since the 1980s, the opportunity to protect the research undertaken by Centers has increased considerably: this is particularly the case in relation to biological innovations and upstream research. While the level of protection varies between countries, nonetheless there has been an expansion in the types of things protected by intellectual property. Alongside developments in patent law that facilitated protection for a spectrum of innovations from plants through to genes, there has also been an increase in the number of countries offering plant breeder's rights protection. This has been prompted by the domestic implementation of the *Agreement on Trade-related Aspects of Intellectual Property Rights* (TRIPS), by the proliferation of bilateral trade agreements that set minimum levels of protection, and by developments in biotechnology.[13] The 1980 decision of the United States Supreme Court of *Diamond v Chacrabarty*[14] also played a role in creating global awareness of the possibility of using patents as a way of protecting genetic innovations. While the decision only confirmed longstanding practice in the United States, nonetheless it was still important in that it raised awareness about intellectual property and the potential role it was able to play in protecting biological innovations.

Another change that increased Center involvement with intellectual property was the realization that exclusivity in general, and intellectual property in particular, were not necessarily antithetical to the goals of poverty reduction and

---

13 A very early example of this is found in *TAC Fact-Finding Mission to the International Center of Insect Physiology and Ecology (ICIPE) Nairobi, Kenya: 24-30 April 1989* where in outlining future work of the Center the Director said that 'commencing work in the field of biotechnology (possibly modifying *Bacillus thuringiensis* for greater activity against the tsetse fly). The question of intellectual property will have to be addressed at the outset': Technical Advisory Committee. 1989. *Report of the TAC Fact Finding Mission to the International Centre of Insect Physiology and Ecology (ICIPE)*. Rome: TAC Secretariat, 15.

14 447 U.S. 303, (1980).

food security.[15] This realization has seen Centers grant private sector organizations the exclusive right to use research products in order either to improve the products or to ensure that they are put into use. In some cases, Centers have also made strategic use of intellectual property themselves in order to enhance the impact that their research has on target beneficiaries.

Another factor that opened the way for intellectual property to play a greater role within CGIAR was the decision to change the way that research in CGIAR was funded, organized and managed. While this reform process had a number of consequences, for my purposes the most important was the decision to introduce what are known as the CGIAR Research Programs. The CGIAR Research Programs are complex multi-party partnerships between a variety of public and private sector agents and Centers. In part, the new research programs were developed in response to the realization that it was not possible for individual Centers to respond to all of the issues that needed to be addressed in dealing with global climate change, agriculture, food security and rural poverty. The new research programs represent a change in the way that research is funded and organized within CGIAR. Instead of a scheme whereby funding flows to Centers who, in turn, liaise with external and internal partners, under the new research programs funding is provided for specific research projects run by partnerships involving Centers (including a lead Center), national public research institutions, non-governmental organizations and private sector partners. These project-focused schemes, which bring together a range of organizations, span the research continuum from the delivery of upstream research through to the adoption of research results by end-users. The reform process also transformed loose research coalitions into more formal (legal) networks. In so doing, the reform process not only pushed intellectual property to the foreground,[16] it also required members to clarify the legal position of the

15 This issue was highlighted in a report by the Genetic Resources Policy Committee presented to the 2000 International Center's Week Meeting that responded to a paper by the Rural Advancement Foundation International that was critical of the *Center's Guiding Principles on IP and Genetic Resources*: Rural Advancement Foundation International. 2000. *The Intellectual Property Challenges to Public Agricultural Research and Human Rights and 28 Alternative Initiatives: In Search of Higher Ground*, The Occasional Paper Series Vol. 6(1). At the meeting, the Chair said that the Centers were committed to the production of global public goods and disagreed with the Rural Advancement Foundation International that the CGIAR guidelines compromised 'public goods in favor of corporate interests' and that they 'constitute a violation of human rights': Consultative Group on International Agricultural Research. 2000. *Charting the CGIAR's Future – Reshaping the CGIAR's Organization*, Report of the 11th Meeting of the Genetic Resources Policy Committee, ICW/00/09. Rome: CGIAR Secretariat, 7-8.

16 See CGIAR Consortium Office, 2010. *Briefing Note for Fund Council meeting Nov 1 and 2, 2010: IP Management to Facilitate Sustained Impact of CRP Research*, 1. [Online]. Available at: http://library.cgiar.org/bitstream/handle/10947/2620/IP_Management_to_Facilitate_Sustained_Impact_of_CRP_Research.pdf?sequence=1 [accessed: 7 February

intellectual property that they brought to the projects, as well as the intellectual property generated from the project.

Another factor that increased the pressure to seek intellectual property protection over CGIAR research has been the growing involvement of the private sector that has, in turn, been prompted by an array of factors.[17] One of the most important external drivers has been the decline in public funding for agricultural research at the national level,[18] and the growth in biotechnology.[19]Another factor that played a role in altering the role that intellectual property plays was the decision within the CGIAR to shift the focus of attention away from the development of research results as ends in their own right to a focus on achieving maximum *impact* to advantage the resource-poor, especially smallholder farmers in developing countries.[20] The focus on maximizing the impact of research was reinforced by the realization that to improve food security it was necessary to engage all actors in the value chain from the delivery of upstream research through to the adoption of research results by end-users (an area where the private sector has considerable experience and expertise).

2013] (explaining how the paradigm shift in the 'new' CGIAR brought about by the CGIAR Research Programs increased the importance of intellectual property in CGIAR).

17 It is important to note that it is not strictly speaking the 'private sector', so much as those organizations who have the exploitation of intellectual property as part of their business plan. In some situations, private sector organizations may operate with a public good goal; while 'public' sector organizations – notably universities and stat-based agencies – may see intellectual property as a means of securing financial returns.

18 For example, the decline in public funding for breeding in Australia was a central justification for the introduction of the plant breeders' rights law in Australia in 1987.

19 See, for example, Private Sector Committee. 1997. *Strengthening CGIAR – Private Sector Partnerships in Biotechnology: A Private Sector Committee Perspective on Compelling Issues*, MTM/97/10, 2-3 [Online]. Available at: http://www.worldbank.org/html/cgiar/publications/mtm97/mtm9710.pdf [accessed: 7 February 2013] (citing IP protection as a critical issue for public–private partnerships). More generally, see World Intellectual Property Organization. 2011. *How the Private and Public Sectors Use Intellectual Property to Enhance Agricultural Productivity.* Geneva: WIPO.

20 In light of this the CGIAR *Strategy and Results Framework* redefines international public goods as 'scientific and technical knowledge, agricultural research products and services or research capacities to respond to and anticipate demand, that are essential to improving agricultural productivity and environmental sustainability in the poor regions of the world .. These goods may find global application, contributing to outcomes in many countries, as can arise through commodity-focused research to improve crop or livestock varieties. Alternatively, they may contribute to more local outcomes, adapted with research partners to specific agricultural systems, as can often arise through participatory breeding programs or with research on policy options or natural resource management.' Consultative Group on International Agricultural Research. 2011. *A Strategy Results Framework for the CGIAR: Submission to the CGIAR Funders Forum.* Washington: CGIAR, 43, note 13.

While the presence of private sector interests may have increased the pressure to use intellectual property to recoup investment in research, there is nothing to suggest that Centers are using intellectual property protection as a means to increase revenue. Indeed, one of the consistent and recurring statements made by Centers was that this was not the case. For example, the International Crops Research Institute for the Semi-Arid Tropics (ICRISAT) 2000 policy on intellectual property stated that '[t]he CGIAR does not view the protection of intellectual property as a mechanism for securing financial returns'.[21] This does not mean, however, that there will not be pressure for this to occur in the future (which is one of the criticisms made of allowing intellectual property into the CGIAR system in the first place),[22] nor that it may not be perceived as a problem. It also does not mean that the involvement of the private sector has not already had an impact on research. Indeed, one change that has occurred is that there is now greater awareness of the role that intellectual property potentially plays in research projects, both in relation to background intellectual property as well as the intellectual property generated from projects. The growing involvement of the private sector also increased the pressure for greater clarity of rights, responsibilities and accountability.[23] It also increased the need to guard against the misappropriation of CGIAR intellectual assets by others. The growing role of the private sector also intensified calls for greater transparency in relation to the use of publicly funded research. Another explanation for the growing interest in using intellectual property to protect the results of CGIAR research is to prevent private sector organizations from locking up research that was intended to be freely available. This occurs where organizations take technologies, traits, varieties and so on, that have freely been made available by Centers without restriction and repackage them in such a way that they restrict access. All of this is done without any obligation to share their profits or assure access. In this situation, Centers are able to use intellectual property to prevent this by controlling what happens to their research when it is disseminated.

---

21 International Crops Research Institute for the Semi-Arid Tropics, 2000. *Policy of the ICRISAT on Intellectual Property Rights*, Annex I (*Guiding Principles on Management of Intellectual Property*). [Online]. Available at: http://test1.icrisat.org/ip_management/policy.htm [accessed: 7 February 2013].

22 The *CGIAR Principles on the Management of Intellectual Assets*, approved by the CGIAR Consortium and the Fund Donors and unanimously supported by the Centers in February 2012, allow for the generation of revenue from the management of intellectual assets (see article 7 of Principles on Fees).

23 While the presence of private sector interests may have increased the pressure to use intellectual property to recoup investment in research, there is little to suggest that Centers are using intellectual property protection as a means to increase revenue. This does not mean, however, that it may not occur in the future (which is one of the criticisms made of allowing intellectual property into the CGIAR system in the first place), nor that it may not be perceived as a problem. It also does not mean that the involvement of the private sector has not had an impact on research.

Another way Centers engage with intellectual property is defensively: that is, they interact with intellectual property with the aim of ensuring that access to research is not restricted.[24] One example of this is the strategic placement of CGIAR publications in patent office databases so that they form part of the prior art that is used when patent applications are examined for novelty and inventiveness. In other instances, Centers have been involved in challenging intellectual property protection taken out by third parties that potentially undermines the ability for Centers to meet their policy needs. One of the most well-known examples is CIAT's successful challenge to the controversial patent application for the Enola bean.[25]

Responding to the Challenges Created by Intellectual Property

While many issues remain unsettled, it is clear that intellectual property now plays a much more prominent role within the CGIAR than has hitherto been the case. Given that intellectual property was initially seen as something that was at best alien and at worst a positive hindrance to the goals of food security and poverty reduction, it is not surprising that the process of change has not been without its problems and challenges. One problem that did arise was the question of whether Centers had the requisite capacity and expertise to deal with and manage intellectual property: an issue that arose when discussions about intellectual property were first raised in the early 1980s. Questions were also raised at the time about the preparedness of Centers to deal with intellectual property, and whether Centers could develop the internal processes and procedures needed to manage intellectual property. As well as worrying about whether there were appropriate skill levels within Centers – or at least those Centers that had some involvement with intellectual property – there was also a concern that Centers would deal with intellectual property in a way that was either inconsistent or contradictory: which would have acted as a disincentive for third parties to engage with Centers.

Perhaps the greatest challenge that arose as part of the assimilation of intellectual property within CGIAR was working out whether or not and, if so, how the goals

---

24 See Adams S. and Henson-Apollonio, V. 2002. *Defensive Publishing: A Strategy for Maintaining Intellectual Property as Public Goods*, International Service for Agricultural Research Briefing Paper 53. [Online]. Available at: ftp://ftp.cgiar.org/isnar/publicat/bp-53.pdf [accessed: 7 February 2013].

25 See International Center for Tropical Agriculture. 2008. *US Patent Office Rejects US Company's Patent Protection for Bean Commonly Grown by Latin American Farmers.* [Online]. Available at: http://webapp.ciat.cgiar.org/newsroom/release_31.htm [accessed: 7 February 2013]; Genetic Resources Policy Committee. 2009. *Minutes of the Genetic Resources Policy Committee: 25th Session*, 3. [Online]. Available at: http://www.cgiar.org/www-archive/www.cgiar.org/pdf/grpc_25th_meeting_minutes.pdf [accessed: 7 February 2013].

of food security and poverty reduction could be reconciled with the growing use of intellectual property. At the heart of this challenge was a concern that intellectual property was, in essence, antithetical to the goals of poverty reduction and food security. The tension inherent in this apparent conflict manifests itself in a myriad of ways. One of which is the potential clash between the ideal of open access to the genetic resources held in trust by the Centers and the situation where a third party takes out intellectual property protection over inventions developed from materials accessed from a Center and so undermines and limits access.

One of the advantages of the traditional approach to intellectual property was its simplicity: the idea that intellectual property was alien to the goals and aspirations of CGIAR was relatively easy to uphold. In part, this was because under the traditional approach, there was no need to balance conflicting goals, no need to justify or explain when and why a Center may have felt it necessary to abrogate from common practice, no concern that this may have undermined the goodwill and trust that external bodies (particularly funding bodies) placed in the Centers and the CGIAR system more generally. The shift away from the traditional approach towards intellectual property has changed this situation. Indeed, one of the most problematic consequences of the move away from the bright line test of the traditional approach was that it gave rise to a series of new questions: in what circumstances is it appropriate for Centers to engage with intellectual property? What could be done to ensure that intellectual property protection does not undermine or distort the core goals of poverty reduction and food security? How do you reconcile the potential dilemma between CGIAR's commitment to providing global access to intellectual assets and some of the Centers' research partner's interests in protecting intellectual property? How do you reconcile the goal of producing goods for which open access can be guaranteed for potential users with the grant of intellectual property rights, which are by their very nature exclusive? How do you ensure that the process of managing intellectual property does not overly burden researchers and managers of Centers? Was the increased use of intellectual property simply a way of channeling money intended for the public good to the private sector?

Over the last decade or so, there has been a concerted effort to respond to these challenges. One area where there has been considerable improvement is in relation to the level of expertise and experience within the Centers and the CGIAR system more generally in relation to intellectual property. Most (if not all) of the Centers now have intellectual property policies; many have dedicated intellectual property managers; there is a broad portfolio of relevant precedents; and a network of people that share ideas and experience. The sharing of templates, precedents and knowledge between Centers has reduced the opportunity costs associated with intellectual property. These developments have been reinforced by the efforts of the CGIAR Consortium Office who have made a conscious effort not only to improve their own expertise in intellectual property law,[26] but also to improve capacity and management skills within the Centers. While there are areas that need improving, there is now a wealth of knowledge and experience about intellectual property within CGIAR.

Another important technique that has been used to respond to the challenges that have accompanied the assimilation of intellectual property in CGIAR is found within the SMTAs that are used whenever genetic resources within the Treaty's Multilateral System are transferred by Centers to third parties. In the same way in which patent law draws a distinction between nature (unimproved and non-patentable) and natural inventions (improved and patentable), so too the MTAs work on the basis that there is a conceptual distinction between unimproved genetic material, which is unable to be patented, and improved genetic material, which is able to be patented. This conceptual distinction is used to ensure that the genetic material held in trust by the Centers is not subject to intellectual property protection, whereas improvements based on that material is. Despite the pivotal role that this conceptual distinction plays, there are many unsettled questions about its scope and operation. This uncertainty is compounded by the fact that although there might be a difference between unimproved and improved material, nonetheless the MTAs also operate on the basis that there is still enough of a connection between the improved and unimproved material (whether legal or genetic) to impose an obligation on third party recipients who improve and commercialize material received from a Center to pay a percentage of the income that they derive from the improved genetic resource.

The third response to the problems posed by the assimilation of intellectual property within CGIAR, which is probably the most innovative and comprehensive, took place in February 2012 when the *CGIAR Principles on the Management of Intellectual Assets* was approved by the CGIAR Consortium and the Fund Donors, unanimously supported by the Centers. These principles form part of the CGIAR Common Operational Framework – which means they apply to everyone working towards the implementation of CGIAR's Strategy and Results Framework. The *CGIAR Principles on the Management of Intellectual Assets* apply to the CGIAR Consortium, its 15 member Centers, and to all their activities and programs. The focus of the Principles is the 'intellectual assets' that are produced or acquired by Centers; that is, the results and/or products of research and development activities that are produced or acquired by Centers. This includes, but is not limited to, knowledge, publications, and other information products, databases, improved germplasm, technologies, inventions, know-how, processes, software and distinctive signs), whether or not they are protected by intellectual property rights.

While the Principles confirm that the ultimate goal of *all* CGIAR activities is the optimal impact on global poverty and food security, they also recognize that in some cases the method that is used to achieve this goal has changed.

26 The Centers established a centralized intellectual property office in the 2000s (known as CAS-IP), which operated for 10 years.

To this end, the Principles aim to provide guidance about how Centers should manage research results and products in this new situation. Specifically, they provide guidance about when access to research results can be restricted (either by intellectual property, contract or other means). The Principles also aim to ensure that the management of intellectual assets is clear, consistent and concise. To allay fears that the CGIAR's engagement with intellectual property specifically and the private sector more generally threatened to undermine the goals of food security and poverty reduction, the Principles aim to ensure that the management of intellectual assets is as transparent and open as possible and, as a result, that Centers are accountable for the way that they spend public funds. The Principles also aim to accommodate differences between Centers, respect Center autonomy and, at the same time, allow CGIAR to develop and exploit the benefits of system-wide approaches on important issues.

One of the key features of the Principles is that they confirm that the ultimate goal of CGIAR is the optimal impact on food security and global poverty. To this end, the Principles provide that Centers are under an obligation to manage the intellectual assets that they produce or acquire so as to maximize their accessibility and to ensure that they have the broadest possible impact on target beneficiaries. Centers are also under a general obligation to manage their research results and products with integrity, fairness, equity, responsibility and accountability. They are also under an obligation to comply with the *Plant Treaty*, the *Convention on Biological Diversity* and relevant human rights conventions.

At the same time, however, the Principles also recognize that in some situations, the goals of food security and poverty reduction can only be achieved by limiting access to CGIAR research. To this end, the Principles set out a number of exceptional situations where Centers are able to limit or restrict access to research results.

In order to enable critical public–private partnerships that benefit smallholder farmers, the Principles recognizes that it may be necessary to grant limited exclusive rights to develop and improve products. The Principles also recognize that it may be necessary to grant exclusive rights to ensure that research is taken up and adopted by end-users. These are exceptional circumstances, and even in these exceptional cases the output would still be available for emergency use and public research in all countries. This exception might be invoked, for example, when a Center develops a new variety, but does not have the ability to distribute it without the assistance of a private seed company that demands an exclusive license for a limited period of time.

The second situation where the Principles allow Centers to restrict access to their research results is when it is necessary to obtain access to third party products and services. It is common practice for Centers to make use of products and processes that are protected by intellectual property. In some situations, the owner of the intellectual property may only allow the Center to use the product on the condition that the Center restricts access to the resulting product. The Principles recognize this and in exceptional circumstances allow Centers to enter into agreements for the acquisition and use of third party products or services that restrict global accessibility to CGIAR products or services. This is on the condition that the restrictions are as limited as possible and the resulting product or service improves food security.

The third exception in the Principles relates to patent and plant variety rights protection. The Principles allow Centers to take out patent and plant variety rights protection over their intellectual assets. The Principles provide that Centers can also allow third parties to take out patent and plant variety rights protection over Center intellectual assets. This is on the condition that the patent or plant variety rights protection is needed either to improve the intellectual assets or to enhance the scale or scope of the impact on target beneficiaries. One of the defining characteristics of patent and plant variety rights protection is that it enables the owner to limit access to the things that are protected. Given this, Centers do not usually seek plant variety protection or patent protection on the products of their research. They do, however, sometimes allow others to do so with regards to products derived from Center lines (i.e. not the Center lines themselves). This would be subject to the terms of the standard material transfer agreement under the *Plant Treaty* which requires payment into a fund in the event the products that are commercialized are restricted for further research and breeding. Centers view the registration by partners of varieties derived from their lines as an important means of promoting the access and benefit sharing principles of the International Treaty.

The fourth and final way in which Centers are able to limit access is by charging fees to access intellectual assets. In most cases, Centers will provide access to their intellectual assets free of charge. In certain limited situations, the Principles allow Centers to charge reasonable financial fees, beyond actual costs and reasonable processing fees, in return for providing access to their respective intellectual assets. This does not apply to the genetic resources held in-trust by Centers covered by the *Plant Treaty.*

As well as outlining how intellectual assets should be managed, the Principles also aim to ensure that the management of intellectual assets is transparent and accountable with respect to the use of public sector funds and activities financed by public sector funds. To promote transparency, Centers must ensure that all policies, guidelines and practices adopted by them are consistent with the Principles and publicly available. Centers are also obliged to provide an annual report to the CGIAR Consortium Office that sets out whether or not and, if so, why they have limited access to intellectual assets. This confidential report, which will be reviewed by the CGIAR Fund Council Intellectual Property Group and followed up on if necessary, is designed to ensure that Centers comply with the Principles and, in so doing, that the goals of CGIAR are not being distorted either by their involvement with IP or their partnerships with private sector organizations.

Conclusion

The last decade or so has seen a series of important changes within CGIAR, one of the most notable being the gradual transformation in the role of IP. These changes are symptomatic of broader changes that are occurring not only in relation to publicly funded agricultural research, but also in adjacent areas such as health.

Typically, intellectual property is perceived as being largely alien to public good research. Under this traditional way of thinking, intellectual property and public good research are diametrically opposed to each other. This is because, on the one hand, the goal of publicly funded research is the generation of research results which are freely available; they are, in legal terms, part of the public domain. On the other hand, while intellectual property law recognizes the public domain, it is fundamentally opposed to free and open access. This dichotomy, so familiar in law, sets up what at first sight is essentially an unresolvable tension. One of the notable things that has occurred over the last decade or so within CGIAR is the innovative way in which this tension has been (or more accurately is being) resolved. What has happened within CGIAR, in effect, is that this problematic dichotomy has been unpacked. This has been done in two ways. The first change occurred when the focus of attention shifted away from research results as ends in their own right towards maximizing the impact of that research. In effect, the cult of the public domain was replaced by a desire to maximize the positive impact that research has on end-users. At the same time, intellectual property was also reconceptualied so that instead of being seen as something that primarily functioned to promote and serve commercial interests, as something that could only protect public interests negatively,[27] it also came to be seen as something that did have something valuable to offer in achieving goals such as reducing poverty and improving food security. While opening up a space in which intellectual property is able operate to improve food security created a number of problems, CGIAR has developed a number of mechanisms that attempt to negate or minimize the negative consequences that often accompany intellectual property. Although it may be some time before we can judge how successful these have been, it is clear that a reconceptualized intellectual property law has an important role to play in helping to improve food security and to reduce poverty and hunger.

27 For example, by limiting protection to 20 years.

Chapter 4

Intellectual Property Norm Setting in *ex situ* Plant Germplasm Access and Benefit Sharing Arrangements

Charles Lawson

Introduction

The preference for cheap food of uniform and predictable quality, the increasing urbanization and move away from subsistence farming, and the increased global population necessitate a huge increase in global agricultural production to feed, clothe and sustain future populations.[1] Added to this are the likely detrimental impacts of climate change.[2] The solution is an increasing reliance on the positive potential for advances in science and technology to increase the efficiency, effectiveness and quantity of food, feed and fibre production. This will require, in part, new plant varieties and access to the germplasm to breed these new varieties. As a direct consequence, these trends and developments have precipitated the need more carefully to conserve and share plant germplasm.

IP is embedded in accessing and benefit sharing in both the germplasm and the incidental institutions and facilities for its conservation and sharing. While patents and plant breeders' rights are the major forms of IP directly affecting the germplasm, other forms of IP play important roles. These include trade secrets to protect lines for hybrids, geographical indications to protect qualities, origins, reputations and characters, copyright over databases and publications and so on. These incidents of IP have both positive and negative consequences, encouraging and rewarding creations and inventions meanwhile also limiting access and increasing prices. The challenge is to understand how IP operates in germplasm exchanges. And then to find an appropriate balance between the needs of inventors,

---

1 Organization for Economic Co-operation and Development and Food and Agriculture Organization of the United Nations. 2012. *OECD–FAO Agricultural Outlook 2012*. Paris: OECD Publishing, 49-81.

2 Intergovernmental Panel on Climate Change. 2012. *Managing the Risks of Extreme Events and Disasters to Advance Climate Change Adaptation. A Special Report of Working Groups I and II of the Intergovernmental Panel on Climate Change*. New York: Cambridge University Press, 246-247.

creators, investors and consumers, and a broader public interest in efficient and effective conservation, exchange and use.

Perhaps the most significant consequence of IP has been its imposition on the previously free sharing of germplasm that characterized the Green Revolution in the 1950s, 1960s and 1970s. With the advent of the United Nations' *Convention on Biological Diversity* (CBD) in 1993 and the Food and Agriculture Organization of the United Nations (FAO) *International Treaty on Plant Genetic Resources for Food and Agriculture* (*Plant Treaty*) in 2004, germplasm is now subject to a regulatory regime that promotes IP in the context of formal agreements (contracts) between the genetic resource holders and those accessing the genetic resource. While the forms of IP are set down in both international and domestic regulation, such as the minimum standard for copyright, patents, trademarks, and so on, in the World Trade Organization's *Agreement on Trade-related Aspects of Intellectual Property Rights* (TRIPS), those implementing the CBD and *Plant Treaty* are establishing norms for the circumstances in which IP will be adopted and asserted. As a consequence the policies and practices of some of those implementing the CBD and *Plant Treaty* are considered in this chapter to investigate the likely contours of IP in practice.

Thus, this chapter reviews the evolving legal architecture implementing the CBD and *Plant Treaty* in the context of *ex situ* germplasm collections (principally the Consultative Group of International Agricultural Research's (CGIAR) germplasm collections) to identify the sources of IP norms affecting germplasm. The chapter demonstrates there are a range of sources and that together these sources are establishing norms for the roles of IP in germplasm exchange arrangements. This is important because accessing the germplasm and then using the materials to improve the food, feed and fibre is the first step in increasing the efficiency and effectiveness of production to deal with necessary increased global agricultural production. The chapter is structured as follows: the next part traces the evolution of the legal architecture that has led to the predominant arrangements for *ex situ* collections under the FAO's *Global System on Plant Genetic Resources for Food and Agriculture*. While this legal architecture does not provide the only means of regulating *ex situ* collections, it does track the evolving concerns about establishing and maintaining *ex situ* collections and the global coordination of global agriculture. The following part outlines both the CBD and the *Plant Treaty* and their IP obligations. The next part then outlines the institution implementing these agreements to demonstrate the sources of authority establishing norms for the ways in which IP is being used in accessing and benefit sharing germplasm exchanges. The final part concludes that there is the potential for IP to affect both the price and availability of food, feed and fibre in the future, although whether this will be problematic is still a question for the future.

The Evolving Legal Architecture

The United Nations *Conference on the Human Environment* in 1972[3] recommended for plants that 'Governments should make sure that valuable gene pools held by individuals or small institutes are also held in national or regional collections',[4] and for 'plant germplasms (agriculture and forestry), organize and equip national or regional genetic resources conservation centers',[5] and that 'the need for liaison among the parties participating in the global system of genetic resources conservation requires certain institutional innovations'.[6] To this end the conference endorsed a recommendation '[t]o assist in the liaison and cooperation between national and regional centers, with special emphasis on international agreements on methodology and standards of conservation of genetic material, standardization and coordination of computerized record systems, and the exchange of information and material between such centers.'[7]

In effect, this was recognition of a coordinating role for the FAO as an impetus for conserving plant germplasm through *ex situ* collections.[8] Around the same time the CGIAR Technical Advisory Committee considered a report by an expert working group about the conservation of the world's genetic resources that had also recommended the development of a global network of germplasm collections and creating a coordination centre.[9] The coordination centre was later established as the International Board for Plant Genetic Resources (IBPGR).[10] The FAO

3 For an overview of the earlier developments see, for example, Pistorious, R. 1997. *Scientists, Plants and Politics: A History of the Plant Genetic Resources Movement*. Rome: International Plant Genetic Resources Institute, 1-33.

4 United Nations General Assembly. 1972. *Report of the United Nations Conference on the Human Environment*, A/CONF.48/14. New York: United Nations, 13 (Recommendation 41(d)(iii)). See also United Nations General Assembly, *United Nations Conference on the Human Environment*, Resolution 2994 (XXVII), 2112th Plenary Meeting 15 December 1972.

5 A/CONF.48/14, 14 (Recommendation 43(2)).

6 Ibid., 15 (Recommendation 45(2)).

7 Ibid., 15 (Recommendation 45(2)(a)(ii)).

8 Notably, the Conference recommended that wild relatives of crop species be maintained in situ through the UNESCO Man and the Biosphere program: ibid., 14 (Recommendation 43(3)).

9 Consultative Group of International Agricultural Research. 1972. *Report of the Third Meeting of the Technical Advisory Committee*, IAR/72/12. Rome: Consultative Group of International Agricultural Research, 21-25. See also Pistorious, above n 3, 48-57.

10 See Baum, W. 1988. *CGIAR – How It All Began*. Rome: Consultative Group of International Agricultural Research, 1-8. Notably, in 1991 IBPGR became the International Plant Genetic Resources Institute (IPGRI), in 1994 IPGRI became a CGIAR Center, and in 2006 IPGRI merged with the International Network for the Improvement of Banana and Plantain (INIBAP) to become a single organization that subsequently changed their operating name to Bioversity International as one of the CGIAR Centers.

and IBPGR then worked together 'promoting the international collaboration of national, regional and international plant genetic centers in which plant genetic resources are collected, maintained, evaluated, exchanged and distributed'.[11] This eventually crystallized as impetus for an international instrument dealing with germplasm together with developing the global network of germplasm collections and an evolving legal architecture for their regulation.[12]

The foundation for the FAO's formal authority for regulating plant genetic resources was the adoption in 1983 of the *International Undertaking on Plant Genetic Resources* (*International Undertaking*)[13] and the establishment of the Commission on Plant Genetic Resources[14] (now the Commission on Genetic Resources for Food and Agriculture – CGRFA – and now with a broadened mandate to cover all components of biodiversity of relevance to food and agriculture).[15] The *International Undertaking* contemplated an international arrangement consisting of a number of centres covering global genetic resources and the necessary expertise to conserve the germplasm.[16] The FAO also established an intergovernmental body to oversee these developments, the now CGRFA.[17]

---

11 Food and Agriculture Organization of the United Nations. 1981. *Report of the Conference of FAO*, Twenty-first Session. Rome: Food and Agriculture Organization, [153] (Resolution 6/81; Preamble). See also Food and Agriculture Organization of the United Nations. 1985. *Report of the Conference of FAO*, Twenty-third Session, C 1985. Rome: Food and Agriculture Organization, [297]. See further Pistorious, above n 3, 55-68; Baum, W. 1986. *Partners Against Hunger: The Consultative Group on International Agricultural Research*. New York: World Bank, 79-82.

12 See Food and Agriculture Organization of the United Nations (Conference, 1981), above n 11, [153] (Resolution 6/81); Food and Agriculture Organization of the United Nations. 1981. *Report of the Council of FAO*, Seventy-ninth Session, CL 79/REP. Rome: Food and Agriculture Organization, [38]. See also Pistorious, above n 3, 79-83.

13 Food and Agriculture Organization of the United Nations. 1983. *Report of the Conference of FAO*, Twenty-second Session. Rome: Food and Agriculture Organization, [285] (Resolution 8/83).

14 Food and Agriculture Organization of the United Nations. 1983. *Report of the Council of FAO*, Eighty-fifth Session, CL 85/REP. Rome: Food and Agriculture Organization, [15] (Resolution 1/85). See also Food and Agriculture Organization of the United Nations (Conference, 1983), ibid., [287] (Resolution 9/83).

15 Food and Agriculture Organization of the United Nations. 1995. *Report of the Conference of FAO*, Twenty-eighth Session. Rome: Food and Agriculture Organization, [69] (Resolution 3/95). See also Food and Agriculture Organization of the United Nations. 1995. *Report of the Council of FAO*, One Hundred and Eighth Session, CL 108/REP. Rome: Food and Agriculture Organization, [15].

16 Food and Agriculture Organization of the United Nations (Conference, 1983), above n 13, [285] (Resolution 8/83; Art 7.1).

17 Food and Agriculture Organization of the United Nations (Conference, 1983), above n 13, [285] (Resolution 8/83; Art 9.2).

The *International Undertaking* was subsequently amended in 1989 to accommodate evolving concerns about IP and biotechnology[18] and farmers' rights,[19] and in 1991 to recognize national sovereignty over genetic resources.[20] Then in 1992 an agreed text of the CBD was settled with express recognition that genetic resources not covered by the CBD needed to be addressed and farmers' rights needed to be resolved.[21] In 1993 the FAO itself resolved to revise the *International Undertaking* in the CGRFA forum in harmony with the CBD.[22]

Negotiations in the CGRFA continued throughout the 1990s, leading eventually to the adoption of the *Plant Treaty* in 2001.[23] Along the way the FAO endorsed the *Global Plan of Action for the Conservation and Sustainable Utilization of Plant Genetic Resources for Food and Agriculture* in 1996 as an internationally agreed framework for the conservation, exploration, collection, characterization, evaluation and documentation of plant genetic resources.[24] And

18 Food and Agriculture Organization of the United Nations. 1989. *Report of the Conference of FAO*, Twenty-fifth Session. Rome: Food and Agriculture Organization, [108] (Resolution 4/89).

19 Ibid., [108] (Resolution 5/89).

20 Food and Agriculture Organization of the United Nations. 1991. *Report of the Conference of FAO*, Twenty-sixth Session. Rome: Food and Agriculture Organization, [104] (Resolution 3/91).

21 Resolution 3 of the Nairobi Final Act: '*Further recognizes* the need to seek solutions to outstanding matters concerning plant genetic resources within the Global System for the Conservation and Sustainable Use of Plant Genetic Resources for Food and Sustainable Agriculture, in particular: (a) Access to ex-situ collections not acquired in accordance with this Convention; and (b) The question of farmers' rights': see Secretariat of the Convention on Biological Diversity. 2005. *Handbook of the Convention on Biological Diversity Including its Cartagena Protocol on Biosafety*, 3rd edition. Montreal: Secretariat of the Convention on Biological Diversity, 408.

22 Food and Agriculture Organization of the United Nations. 1993. *Report of the Conference of FAO*, Twenty-seventh Session. Rome: Food and Agriculture Organization, [105] (Resolution 7/93). See also Conference of the Parties to the Convention on Biological Diversity. 1994. *Medium-term Program of Work of the Conference of the Parties: Progress Report by the Food and Agriculture Organization of the United Nations (FAO) on work being carried out by the Commission on Plant Genetic Resources to implement Resolution 3 of the Nairobi Final Act*, UNEP/CBD/COP/1/INF/3. Montreal: Secretariat of the Convention on Biological Diversity, [9]-[12].

23 Food and Agriculture Organization of the United Nations. 2001. *Report of the Conference of FAO*, Thirty-first Session, C 2001/REP. Rome: Food and Agriculture Organization, [58] (Resolution 3/2001).

24 Food and Agriculture Organization of the United Nations. 1997. *Report of the Conference of FAO*, Twenty-ninth Session. Rome: Food and Agriculture Organization, [48] (Resolution 1/97). See also Fourth International Technical Conference on Plant Genetic Resources. 1996. *Report of the International Technical Conference on Plant Genetic Resources*, ITCPGR/96/REP. Rome: Food and Agriculture Organization; Conference of the Parties to the Convention on Biological Diversity. 2000. *Report of the Fifth Meeting*

in 2000, the Conference of the Parties to the CBD 'envisaged' that the revised *International Undertaking* (the now *Plant Treaty*) would 'play a crucial role in the implementation of the [CBD]',[25] and that the *Plant Treaty* system was integral to national responses to implement the CBD.[26] These various influences are expressly recognized in the *Plant Treaty* providing, in part, in Art 1:[27]

> The objectives of this [*Plant Treaty*] are the conservation and sustainable use of plant genetic resources for food and agriculture and the fair and equitable sharing of the benefits arising out of their use, in harmony with the *Convention on Biological Diversity* ... These objectives will be attained by closely linking this [*Plant Treaty*] to the Food and Agriculture Organization of the United Nations and to the *Convention on Biological Diversity*.

And also, in part, in Art 14: 'Recognizing that the rolling [*Global Plan of Action*] is important to this [*Plant Treaty*], Contracting Parties should promote its effective implementation, including through national actions and, as appropriate, international cooperation to provide a coherent framework, inter alia, for capacity-building, technology transfer and exchange of information ...'.

In 2006 the *Plant Treaty*'s Governing Body determined that the *Global Plan of Action*'s priorities were also priorities for the *Plant Treaty*'s Funding Strategy.[28] Then in 2009 the *Plant Treaty*'s Governing Body invited the CGRFA, when revising the *Global Plan of Action*, to take into account the *Plant Treaty* and reflect the *Plant Treaty*'s provisions in the revised *Global Plan of Action*.[29] The effect of the *Plant Treaty* has been to link the various developments affecting IP in the *Plant Treaty*, the FAO, the CBD and the *Global Plan of Action*. This legal architecture is referred to collectively as the FAO's *Global System on Plant Genetic*

---

of the Conference of the Parties to the Convention on Biological Diversity, UNEP/CBD/COP/5/23. Montreal: Secretariat of the Convention on Biological Diversity, 123 ([2]; Decision V/10)).

25 UNEP/CBD/COP/5/23, 86 ([12]; Decision V/5)).

26 Ibid., 197 ([7]-[8]; Decision V/26A)).

27 See also *International Treaty on Plant Genetic Resources for Food and Agriculture*, Arts 15.3, 17.1, 19.3(g), (l) and (m), and 20.5.

28 Governing Body of the International Treaty on Plant Genetic Resources for Food and Agriculture. 2006. *Report of the First Session of the Governing Body of the International Treaty on Plant Genetic Resources for Food and Agriculture*, IT/GB-1/06/REPORT. Rome: Governing Body, [11] (Resolution 1/2006).

29 Governing Body of the International Treaty on Plant Genetic Resources for Food and Agriculture. 2009. *Report of the Governing Body of the International Treaty on Plant Genetic Resources for Food and Agriculture*, IT/GB-3/09/Report. Rome: Governing Body, [50] and Appendix A.7 (Resolution 7/2009).

Resources for Food and Agriculture.[30] Reflecting the balance between the different institutions and their capacities to deliver on various aspects of the conservation and sustainable uses of genetic resources, these developments have resulted in a collection of institutions, agreements, mechanisms and instruments directed to conserving genetic resources for food and agriculture. This now comprises (monitored by the CGRFA and its Intergovernmental Technical Working Group on Plant Genetic Resources for Food and Agriculture):[31]

a. *Institutions* – The key institution is the CGRFA that provides a forum for donors or users of germplasm, funds and technology to discuss, on an equal footing, matters related to plant genetic resources.[32] To further assist the CGRFA there is the Intergovernmental Technical Working Group on Plant Genetic Resources for Food and Agriculture.[33] Following the adoption of the *Plant Treaty* the Governing Body also has a key role to 'provide policy direction and guidance to monitor and adopt such recommendations as necessary for the implementation of the [*Plant Treaty*] and, in particular, for the operation of the Multilateral System'.[34]
b. *Agreements* – The relevant agreements included *Plant Treaty*,[35] the *International Code of Conduct for Plant Germplasm Collecting and Transfer*,[36] draft *Code of Conduct on Biotechnology*,[37] and various gene bank standards and guidelines.[38]

---

30 See Commission on Plant Genetic Resources. 1995. *Progress Report on the Global System for the Conservation and Utilization of Plant Genetic Resources for Food and Agriculture*, CPGR-6/95/4, Rome: Commission on Plant Genetic Resources, [5]-[63].

31 See Commission on Plant Genetic Resources. 1997. *Progress Report on the Commission on Plant Genetic Resources*, CGRFA-7/97/REP. Rome: Commission on Plant Genetic Resources, [44] and Appendix F; Food and Agriculture Organization of the United Nations (Conference, 1983), above n 13, [285] (Resolution 8/83; Art 9.2).

32 Food and Agriculture Organization of the United Nations (Conference, 1983), above n 13, [287] (Resolution 9/83). See also Food and Agriculture Organization of the United Nations (Council, 1983), above n 14, [15] (Resolution 1/85).

33 See CGRFA-7/97/REP, above n 31, [10], [44] and Appendix F.

34 *International Treaty on Plant Genetic Resources for Food and Agriculture*, Art 19.3(a).

35 Food and Agriculture Organization of the United Nations (Conference, 2001), above n 23, [58] (Resolution 3/2001).

36 Food and Agriculture Organization of the United Nations (Conference, 1993), above n 22, [113] and Appendix E (Resolution 8/93).

37 See Commission on Genetic Resources for Food and Agriculture. 2004. *Progress in the Draft Code of Conduct on Biotechnology as it relates to Genetic Resources for Food and Agriculture: Policy Issues, Gaps and Duplications*, CGRFA-10/04/13. Rome: Commission on Genetic Resources for Food and Agriculture.

38 Commission on Plant Genetic Resources. 1993. *Report of the Commission on Plant Genetic Resources*, CPGR/93/REP. Rome: Commission on Plant Genetic Resources,

c. *Mechanisms* – World Information and Early Warning System of Plant Genetic Resources for Food and Agriculture (WIEWS),[39] crop and thematic networks (facilitation mechanisms),[40] and the international network of *ex situ* collections (including the CGIAR Centers under the *Plant Treaty* Article 15 agreements) and *in situ* on-farm areas.[41]

d. *Instruments* – The relevant instruments include the *State of the World's Plant Genetic Resources for Food and Agriculture*,[42] the *Second Report on the State of the World's Plant Genetic Resources for Food and Agriculture*,[43] the *Global Plan of Action for the Conservation and Sustainable Utilization of Plant Genetic Resources for Food and Agriculture*,[44] and the *Second Global Plan of Action for Plant Genetic Resources for Food and Agriculture*.[45]

---

[30]. Notably these standards and guidelines are under review: see also Governing Body of the International Treaty on Plant Genetic Resources for Food and Agriculture. 2009. *Twelfth Regular Session of the Commission on Genetic Resources for Food and Agriculture*, CGRFA-12/09/Report. Rome: Governing Body, [89]; IT/GB-3/09/Report, above n 29, [50] and Appendix A.7 (Resolution 7/2009).

39 This is essentially a collection of databases: a country profiles database; an *ex situ* collection database; a laws and regulations database; the *World List of Seed Sources* database; and the *List of Crop Varieties* database: see Commission on Genetic Resources for Food and Agriculture. 2002. *Progress Report on the World Information and Early Warning System on Plant Genetic Resources for Food and Agriculture*, CGRFA-9/02/10. Rome: Commission on Genetic Resources for Food and Agriculture.

40 A range of activities in conservation and sustainable utilization of plant genetic resources for food and agriculture are managed by local, regional and international institutions and network that are loosely identified by their association with the *Global Plan of Action*: see http://www.globalplanofaction.org. See also Commission on Genetic Resources for Food and Agriculture. 2007. *Follow Up to Recommendations of the Commission on Genetic Resources for Food and Agriculture Regarding Plant Genetic Resources for Food and Agriculture*, CGRFA-11/07/11. Rome: Commission on Genetic Resources for Food and Agriculture, [7] and Appendix.

41 See Food and Agriculture Organization of the United Nations. 2010. *Second Report on the State of the World's Plant Genetic Resources for Food and Agriculture*. Rome: Food and Agriculture Organization, 31-46 (*in situ*) and 55-88 (*ex situ*).

42 Food and Agriculture Organization of the United Nations. 1996. *Report on the State of the World's Plant Genetic Resources for Food and Agriculture*. Rome: Food and Agriculture Organization. See also Food and Agriculture Organization of the United Nations (Conference, 1997), above n 24, [48] (Resolution 1/97); CGRFA-7/97/REP, above n 31, [12]; ITCPGR/96/REP, above n 24, [14].

43 *Second Report*, above n 41; Commission on Genetic Resources for Food and Agriculture. 2009. *Draft Second Report on The State of the World's Plant Genetic Resources for Food and Agriculture - Final Version*, CGRFA-12/09/Inf.7 Rev.1. Rome: Commission on Genetic Resources for Food and Agriculture. See also CGRFA-12/09/Report, above n 38, [17].

44 ITCPGR/96/REP, above n 24, [15] and Annex 2.

45 Food and Agriculture Organization of the United Nations. 2011. *Report of the Council of FAO*, CL 143/REP. Rome: Food and Agriculture Organization, [43]. See also

The key agreements are considered next.

Threshold Agreements

The uppermost levels of formal regulation are the international agreements. In dealing with accessing germplasm these include a general scheme under the United Nations' CBD and a specific scheme dealing with a limited number of agriculturally significant plants under the FAO's *Plant Treaty*. While each of these schemes does not establish minimum standard IP, both schemes establish norms for the ways in which IP is addressed.

Convention on Biological Diversity (CBD)

The CBD provides a scheme for access to 'genetic resources'[46] according to the authority of the Nation State '[r]ecognising the sovereign rights of States over their natural resources'[47] with an obligation to facilitate access for 'environmental sound uses' without imposing restrictions that are counter to the CBD's objectives.[48] Further, access must be from countries of origin or countries that have acquired the genetic resources according to the CBD,[49] on mutually agreed terms,[50] with prior informed consent,[51] and most importantly, taking:

> legislative, administrative or policy measures, as appropriate, and in accordance with Arts 16 [access to and transfer of technology] and 19 [handling of biotechnology and distribution of its benefits] and where necessary through

---

Commission on Genetic Resources for Food and Agriculture. 2011. *Report of the 13th Regular Session of the Commission on Genetic Resources for Food and Agriculture*, CGRFA-13/11/Report. Rome: Commission on Genetic Resources for Food and Agriculture, [21] and Appendix B (Resolution 1/2011 (Annex: *Second Global Plan of Action for Plant Genetic Resources for Food and Agriculture*)).

46 'Genetic resources' is broadly defined to mean 'genetic material of actual or potential value' and 'genetic materials' means 'any material of plant, animal, microbial or other origin containing functional units of heredity': *Convention on Biological Diversity*, Art 2. Notably, definition includes biochemicals, excludes *ex-situ* holdings acquired before 29 December 1993, excludes human genetic materials, and excludes marine living (genetic and biochemical) resources beyond national jurisdiction: Conference of the Parties to the Convention on Biological Diversity. 1995. *Access to Genetic Resources and Benefit-sharing: Legislation, Administrative and Policy Information*, UNEP/CBD/COP/2/13. Montreal: Secretary to the Convention on Biological Diversity, [49]-[65].

47 *Convention on Biological Diversity*, Art 15(1). See also Art 3.

48 Ibid., Art 15(2).

49 Ibid., Art 15(3).

50 Ibid., Art 15(4).

51 Ibid., Art 15(5).

> the financial mechanism established by Arts 20 [financial resources] and 21 [financial mechanism] with the aim of sharing in a fair and equitable way the results of research and development and the benefits arising from the commercial and other utilisation of genetic resources with the Contracting Party providing such resources.[52]

In resolving how the CBD might work in practice, the Conference of the Parties to the Convention on Biological Diversity agreed to the *Bonn Guidelines on Access to Genetic Resources and Fair and Equitable Sharing of the Benefits Arising out of their Utilisation* (Bonn Guidelines)[53] as voluntary guidelines that apply to *all* genetic resources covered by the CBD (except human genetic resources),[54] in a manner that is 'coherent and mutually supportive of the work of relevant international agreements and institutions'[55] and 'without prejudice' to the *Plant Treaty*.[56]

This was 'merely the first step on a long and complex process to secure access and benefit-sharing' under the CBD,[57] 'recognizing' that 'a package of measures may be necessary to address the different needs of Parties and stakeholders in the implementation of access and benefit-sharing arrangements'.[58] This was responding to broader concerns about fairly and equitable share the benefits from using genetic resources under the CBD that had been expressed at the UN *World Summit on Sustainable Development*[59] and at the UN General Assembly.[60] The outcome of further negotiations was the *Nagoya Protocol on Access to Genetic Resources and the Fair and Equitable Sharing of Benefits Arising from their Utilization* (Nagoya Protocol).[61]

---

52 Ibid., Art 15(7).

53 Conference of the Parties to the Convention on Biological Diversity. 2002. *Report of the Sixth Meeting of the Conference of the Parties to the Convention on Biological Diversity*, UNEP/CBD/COP/6/20. Montreal: Secretary to the Convention on Biological Diversity, [342] and Annex I (Decision VI/24A, 253-269 (Bonn Guidelines)).

54 UNEP/CBD/COP/6/20, above n 53, [342] and Annex I (Decision VI/24A, [9], 254).

55 Ibid., [342] and Annex I (Decision VI/24A, [10], 254).

56 Ibid., [342] and Annex I (Decision VI/24A, [10], 254).

57 Ibid., [42].

58 Ibid., [342] and Annex I (Decision VI/24B, [10]-[12], 271).

59 See United Nations. 2002. *Report of the World Summit on Sustainable Development*, A/CONF.199/20. New York: United Nations, [44(o)] (Plan of Implementation).

60 See General Assembly of the United Nations. 2003. *Convention on Biological Diversity*, Resolution 57/260, 78th Plenary Meeting, 20 December 2002, A/RES/57/260. New York: United Nations, [7]-[9].

61 Conference of the Parties to the Convention on Biological Diversity. 2010. *Report of the Tenth Meeting of the Conference of the Parties to the Convention on Biological*

The obligations imposed by the Nagoya Protocol are essentially that Members will establish appropriate legislative, administrative or policy measures to ensure that any benefits (monetary and non-monetary) are shared with the originating holder of the resource.[62] Where these are indigenous and local communities, any benefits (monetary and non-monetary) are to be shared with the indigenous and local communities concerned.[63] And further, where 'traditional knowledge associated with genetic resources' are used, then there must be appropriate legislative, administrative or policy measures to ensure that any benefits (monetary and non-monetary) are shared with the 'indigenous and local communities holding such knowledge'.[64] The Nagoya Protocol restates the CBD's requirements of mutually agreeable terms and prior informed consent and extends this to include the relevant 'indigenous and local communities' and 'approval and involvement of indigenous and local communities' respectively.[65] Significantly, the Protocol sets out necessary procedural standards for establishing prior informed consent[66] and requires the establishment of a 'national focal point on access and benefit-sharing'[67] and an 'Access and Benefit-sharing Clearing-House'.[68] The clearing-house is anticipated to provide access to 'model contractual clauses', 'methods and tools developed to monitor genetic resources' and 'codes of conduct and best practices'.[69]

The Nagoya Protocol marked a significant recognition of the existing IP standards by accepting that other international agreements continued to have force (and while not explicitly mentioned, clearly TRIPS was in mind).[70]

The outcome of the CBD, Bonn Guidelines and Nagoya Protocol is that IP is a matter for the parties to determine at the time they are negotiating the access and benefit-sharing arrangements as a term or condition of the access and benefit-sharing contract. In achieving this, the CBD makes general provision that IP as part of access to technology in exchange for access to genetic resources 'shall be provided on terms which recognize and are consistent with the adequate and

Diversity, UNEP/CBD/COP/10/27. Montreal: Secretary to the Convention on Biological Diversity, [103] and Annex (Decision X/1, Annex 1 (Nagoya Protocol), 89-109).

62 UNEP/CBD/COP/10/27, above n 61, [103] and Annex (Decision X/1, Annex 1 (Art 5(1)), 92).

63 Ibid., [103] and Annex (Decision X/1, Annex 1 (Art 5(2)), 92).

64 Ibid., [103] and Annex (Decision X/1, Annex 1 (Art 5(5)), 92).

65 Ibid., [103] and Annex (Decision X/1, Annex 1 (Arts 5(1) and (2), 6(1) and (2) and 7)), 91-94).

66 Ibid., [103] and Annex (Decision X/1, Annex 1 (Art 6(3)), 93).

67 Ibid., [103] and Annex (Decision X/1, Annex 1 (Art 13), 96).

68 Ibid., [103] and Annex (Decision X/1, Annex 1 (Art 14), 96-97).

69 Ibid., [103] and Annex (Decision X/1, Annex 1 (Arts 14(3) and 19 and 20), 97 and 100).

70 See ibid., [103] and Annex (Decision X/1, Annex 1 (Art 4), 91-92).

effective protection of IP rights',[71] and that some measures might be necessary to compensate non-government providers of IP as a part of access and benefit-sharing agreements.[72] The CBD also requires cooperation to ensure IP does not 'run counter' to the CBD's conservation and access and benefit sharing objectives.[73]

The interaction between IP and the CBD remains contentious and unresolved with internationally contested inherent conflicts between TRIPS and the CBD being that TRIPS requires genetic materials be protected by patents or a *sui generis* plant variety that privately appropriates genetic resources over which a country has sovereign rights under the CBD,[74] and that these privileges do not also require the additional measures set out in the CBD, such as prior informed consent, mutually agreed terms and benefit sharing.[75] The CBD issue also remains deadlocked in the main IP forums at the WTO with three apparent blocks: those favouring amending TRIPS, those seeing no conflict between TRIPS and the CBD, and those considering that the WTO's Council of TRIPS is addressing the issue and that it should not be duplicated by the WTO's Committee on Trade and Environment.[76]

The advance of the Nagoya Protocol was to impose requirements that the country of origin of the resources establish rules and procedures for assessing mutually agreed terms (including the terms of benefit sharing addressing IP)[77] by issuing a permit[78] through a competent national authority[79] and information ('[w]ithout prejudice to the protection of confidential information')[80] to an Access

71 *Convention on Biological Diversity*, Art 16(2).

72 Ibid., Art 16(3). See also UN Conference of Environment and Development. 1992. *Report of the United Nations Conference of Environment and Development*, A/CONF.151/26/Rev.l (Vol. 1). New York: United Nations, 422 ([34.18(e)(iii)]).

73 *Convention on Biological Diversity*, Art 16(5).

74 Council for Trade-Related Aspects of Intellectual Property Rights. 2002. *The Relationship Between the TRIPs Agreement and the Convention on Biological Diversity – Summary of Issues Raised and Points Made*, IP/C/W/368. Geneva: World Trade Organization, 2. See also Council for Trade-Related Aspects of Intellectual Property Rights. 1999. *Review of the Provisions of Art 27.3.b - Communication from Kenya on Behalf of the African Group*, IP/C/W/163. Geneva: World Trade Organization.

75 Council for Trade-Related Aspects of Intellectual Property Rights. 2000. *Minutes of Meeting Held in the Centre William Rappard on 21 and 22 September 2000*, IP/C/M/28. Geneva: World Trade Organization, 43.

76 Secretariat of the World Trade Organization. 2004. *Trade and Environment at the WTO*, Geneva: World Trade Organization, 42.

77 UNEP/CBD/COP/10/27, above n 61, [103] and Annex (Decision X/1, Annex 1 (Art 6(3)), 93).

78 Ibid., [103] and Annex (Decision X/1, Annex 1 (Arts 6(3) and 17(2)), 93 and 99).

79 Ibid., [103] and Annex (Decision X/1, Annex 1 (Art 13(2)), 96).

80 Ibid., [103] and Annex (Decision X/1, Annex 1 (Art 14(2)), 97).

and Benefit-sharing Clearing-House mechanism.[81] A further obligation is to 'take appropriate, effective and proportionate legislative, administrative or policy measures' to ensure prior informed consent, mutually agreed terms and traditional knowledge obligations have been complied with[82] through 'checkpoints'.[83] The Access and Benefit-sharing Clearing-House is also intended to provide '[m]odel contractual clauses'[84] and '[c]odes of conduct and best practices'.[85] The Conference of Parties to the CBD provides periodic stock-take of model contract clauses,[86] albeit this has not happened yet. The effect over time will be to establish norms for the ways IP is addressed and the ways IP is structured in access and benefit sharing agreements.

International Treaty on Plant Genetic Resources for Food and Agriculture (Plant Treaty)

The *Plant Treaty* provides a scheme for accessing plant genetic resources forming the 'efficient, effective, and transparent' Multilateral System.[87] The Multilateral System includes 'all plant genetic resources for food and agriculture listed in Annex 1 that are under the management and control of the Contracting Parties and in the public domain',[88] contributions from the 'the *ex situ* collections of the International Agricultural Research Centers (IARC) of the CGIAR'[89] and contributions from 'other international institutions'.[90] Access to these materials in the Multilateral System is according to a 'Standard Material Transfer Agreement' (SMTA) that has been adopted by the Treaty's Governing Body.[91] Other materials collected after 29 June 2004 must comply with the other international obligations, primarily the CBD's prior informed consent and mutually agreed terms (addressed above).

The *Plant Treaty* makes express provision for IP, providing that access to the Multilateral System must comply with the condition that:

---

81 Ibid., [103] and Annex (Decision X/1, Annex 1 (Art 14(1)), 96).

82 Ibid., [103] and Annex (Decision X/1, Annex 1 (Arts 15(1) and 16(1)), 97 and 98).

83 Ibid., [103] and Annex (Decision X/1, Annex 1 (Arts 17(1)), 98-99).

84 Ibid., [103] and Annex (Decision X/1, Annex 1 (Arts 14(3) and 19(1)), 97 and 100).

85 Ibid., [103] and Annex (Decision X/1, Annex 1 (Art 14(3)), 97).

86 Ibid., [103] and Annex (Decision X/1, Annex 1 (Art 19(2)), 100).

87 *International Treaty on Plant Genetic Resources for Food and Agriculture*, Art 10.2.

88 Ibid., Art 11.2.

89 Ibid., Arts 11.5 and 15.1(a).

90 Ibid., Arts 11.5 and 15.5.

91 See IT/GB-1/06/REPORT, above n 28, [12] (Resolution 2/2006) and Appendix G (Standard Material Transfer Agreement).

> Recipients shall not claim any [IP] or other rights that limit the facilitated access to the plant genetic resources for food and agriculture, or their genetic parts or components, in the form received from the Multilateral System.[92]
>
> Access to plant genetic resources for food and agriculture protected by intellectual and other property rights shall be consistent with relevant international agreements, and with relevant national laws.[93]

Further provision is made for access and transfer of technology, including that protected by IP.[94] In short, the *Plant Treaty* provides that IP should not be claimed over materials received by the Recipient under an SMTA, albeit any changes or modifications by the Recipient may be subject to IP claims. Where materials are outside the scope of the *Plant Treaty* and to be used for purposes other than 'the purposes of research, breeding and training for food and agriculture' (excluding 'chemical, pharmaceutical and/or other non-food/feed industrial uses')[95], then IP can be claimed. The meaning of these terms remains uncertain and will likely be clarified through the experience of contracting according to the norms established by the Governing Body (and the CGIAR Centers) for implementing the *Plant Treaty*.

The Institutions Effecting the Agreements

The analysis of the CBD and *Plant Treaty* demonstrate the ways these agreements embrace existing international commitments to IP, such as the TRIPS, and then make provision for IP to be incorporated in access and benefit sharing arrangements. The institutions putting into effect the CBD and the *Plant Treaty* (doing the accessing, conserving and benefit sharing) that will establish norms for these IP obligations. While under the CBD there will be the plethora of national competent authorities reporting through the periodic stock-takes of model contract clauses to the Conference of the Parties to the CBD,[96] this is not readily apparent yet.[97] There are, however, other institutions that have provided some insights and these are considered here.

---

92 *International Treaty on Plant Genetic Resources for Food and Agriculture*, Art 12.3(d).

93 Ibid., Art 12.3(f).

94 Ibid., Arts 13.2(b)(i) and (iii).

95 IT/GB-1/06/REPORT, above n 28, Appendix G (Standard Material Transfer Agreement) cl 6.1.

96 UNEP/CBD/COP/10/27, above n 61, [103] and Annex (Decision X/1, Annex 1 (Art 19(2)), 100).

97 There is some data collected so far and a small sample of some contracts: see World Intellectual Property Organization, *Database of Biodiversity-related Access and*

Commission on Genetic Resources for Food and Agriculture (CGRFA)

The CGRFA was expressly established by the FAO as a part of responding to the *International Undertaking*.[98] Key functions of the CGRFA are now a broad mandate to cover all components of biodiversity of relevance to food and agriculture[99] and to monitor (with the Intergovernmental Technical Working Group on Plant Genetic Resources for Food and Agriculture)[100] the implementation of the *Global Plan of Action*[101] as part of the FAO's *Global System on Plant Genetic Resources for Food and Agriculture*.[102] While IP is not an explicit focus of the CGRFA, it is an issue that impacts on various aspects of their work and an acknowledged matter of interest. This has arisen as a result of the CGRFA's involvement as the FAO's forum for dealing with plant genetic resources and in dealing with the Global Plan of Action. These are addressed in turn.

The FAO in 1989 expressly adopted an *Agreed Interpretation of the International Undertaking* that recognized the role of IP providing, in part:[103]

> Plant Breeders' Rights as provided for under UPOV (International Union for the Protection of New Varieties of Plant) are not incompatible with the International Undertaking … A state may impose only such minimum restriction on the free exchange of materials … as are necessary for it to conform to its national and international obligations.[104]

At the same time the FAO endorsed the concept of farmers' rights being:

> rights arising from the past, present and future contributions of farmers in conserving, improving, and making available plant genetic resources, particularly those in the International Community, as trustee for present and future generations of farmers, for the purpose of ensuring full benefits to farmers,

---

Benefit-sharing Agreements. [Online]. Available at: http://www.wipo.int/tk/en/databases/contracts/list.html [accessed: 3 January 2013].

98 Food and Agriculture Organization of the United Nations (Conference, 1983), above n 13, [287] (Resolution 9/83). See also Food and Agriculture Organization of the United Nations (Council, 1983), above n 14, [15] (Resolution 1/85).

99 Food and Agriculture Organization of the United Nations (1985), above n 11, [69] (Resolution 3/95). See also Food and Agriculture Organization of the United Nations (Council, 1995), above n 15, [15].

100 See CGRFA-7/97/REP, above n 31, [10] and Appendix F.

101 See ibid., [44] and Appendix F; Food and Agriculture Organization of the United Nations (Conference, 1983), above n 13, [285] (Resolution 8/83; Art 9.2).

102 See CPGR-6/95/4, above n 30, [5]-[63].

103 Food and Agriculture Organization of the United Nations (Conference, 1989), above n 18, [108] (Resolution 4/89).

104 Ibid., [108] (Resolution 4/89; Arts 1 and 2).

> and supporting the continuation of their contributions, as well as the attainment of the overall purposes of the *International Undertaking*.[105]

More recently, the CGRFA was the forum for revising the *International Undertaking* in harmony with the CBD, considering the issue of access on mutually agreed terms to plant genetic resources, including some *ex situ* collections and the realization of farmers' rights.[106] The outcome of the revised *International Undertaking* was the *Plant Treaty* and its provision for IP (dealt with above), and express provision recognizing farmers' rights that is now dealt with by the Governing Body of the *Plant Treaty*.[107] There remains, however, an ongoing interest in IP through regularly reviewing developments in all relevant forums as an international cross-sectorial policy issue,[108] a key issue in the *Strategic Plan 2010-2017 for the implementation of the Multi-Year Programme of Work*,[109] and the potential to establish further norms.[110]

The *Global Plan of Action* was adopted in 1996.[111] The initial plan essentially sets out 20 priority activities in four broad groupings: (i) *in situ* conservation and development; (ii) *ex situ* conservation; (iii) the use of plant genetic resources;

105 Ibid., [108] (Resolution 5/89).

106 Food and Agriculture Organization of the United Nations (Conference, 1993), above n 22, [105] (Resolution 7/93; Art 1).

107 Governing Body of the International Treaty on Plant Genetic Resources for Food and Agriculture. 2011. *Report of the Fourth Session of the Governing Body of the International Treaty on Plant Genetic Resources for Food and Agriculture*, IT/GB-4/11/Report. Rome: Governing Body, [29] and Appendix A.6 (Resolution 6/2011). See also Governing Body of the International Treaty on Plant Genetic Resources for Food and Agriculture. 2011. *Implementation of Article 9 Farmers' Rights*, IT/GB-4/11/16. Rome: Governing Body.

108 See CGRFA-12/09/Report, above n 38, [11] and Appendix G ([36]); Commission on Genetic Resources for Food and Agriculture. 2007. *Eleventh Regular Session of the Commission on Genetic Resources for Food and Agriculture*, CGRFA-11/07/Report. Rome: Commission on Genetic Resources for Food and Agriculture, [72]. See also Correa, C. 2009. *Trends in Intellectual Property Rights Relating to Genetic Resources for Food and Agriculture*, Background Study Paper No 49. Rome: Commission on Genetic Resources for Food and Agriculture.

109 See CGRFA-12/09/Report, above n 38, [11] and Appendix G ([36], [39] and Indicative Timetable and Processes (p 29)).

110 See ibid. [37] and [39].

111 ITCPGR/96/REP, above n 24, [15] and Annex 2 (*Global Plan of Action for the Conservation and Sustainable Utilization of Plant Genetic Resources for Food and Agriculture*). See also Food and Agriculture Organization of the United Nations (Conference, 1997), above n 24, [48] (Resolution 1/97); Food and Agriculture Organization of the United Nations. 1996. *Report of the Council of FAO*, Twenty-ninth Session. Rome: Food and Agriculture Organization, [21] (Resolution 1/111).

and (iv) institution and capacity building.[112] The *Global Plan of Action* itself was compiled through a preparatory process involving meetings of countries providing recommendations about activities, priorities and programmes in the area of plant genetic resources of concern and various reports including the *State of the World's Plant Genetic Resources for Food and Agriculture*.[113] The first *Global Plan of Action* adopted in 1996[114] was based on the first *State of the World's Plant Genetic Resources for Food and Agriculture*.[115] Since then the *Global Plan of Action* has been reviewed[116] based on the *Second of the State of the World's Plant Genetic Resources for Food and Agriculture*.[117] The second *Global Plan of Action* details a broad strategic framework to enhance conservation and use of plant genetic resources for food and agriculture through establishing standards for operating germplasm banks, promoting information collections and exchanges, capacity building among contributors to a 'true global system', enhancing linkages, promoting *in situ* conservation, promoting the conservation of crop wild relatives, and promoting '[c]onservation and use at the community, national, regional and international levels'.[118] The second plan details 18 priority activities organized into four main groups: (i) *in situ* conservation and management; (ii) *ex situ* conservation; (iii) sustainable use; and (iv) building sustainable institutional and human capacities.[119]

While IP is not an immediate focus of the *Global Plan of Action*, IP is likely to affect its implementation. This is expressly recognized in the context of building and strengthening national programs:

> The creation or strengthening of links should be promoted, as appropriate, among national institutions and entities specialized in technology transfer, in order to assist national bodies to negotiate the acquisition of technologies needed to conserve, characterize, and use [plant genetic resources for food and

112 See Commission on Genetic Resources for Food and Agriculture. 2009. *Updating the Global Plan of Action for the Conservation and Sustainable Utilization of Plant Genetic Resources for Food and Agriculture*, CGRFA/WG-PGR-4/09/3. Rome: Commission on Genetic Resources for Food and Agriculture, [4].

113 See ibid., [5]-[7].

114 ITCPGR/96/REP, above n 24, [15] and Annex 2.

115 Commission on Genetic Resources for Food and Agriculture. 1997. *Report of the State of the World's Plant Genetic Resources for Food and Agriculture*. Rome: Commission on Genetic Resources for Food and Agriculture.

116 Food and Agriculture Organization of the United Nations (Council, 2011), above n 45, [43]. See also CGRFA-13/11/Report, above n 45, [21] and Appendix B (Resolution 1/2011 (Annex: *Second Global Plan of Action for Plant Genetic Resources for Food and Agriculture*)).

117 *Second Report*, above n 41. See CGRFA-12/09/Report, above n 38, [17].

118 CGRFA-13/11/Report, above n 45, Appendix B (Annex: [19]).

119 Ibid., Appendix B (Annex: [21] and [24]-[312]).

agriculture] and associated information, in accordance with the [*Plant Treaty*], the CBD and [IP] rights.[120]

The CGRFA's *Strategic Plan 2010-2017 for the implementation of the Multi-Year Programme of Work* expressly recognizes that IP is in issue[121] and that any deliberations have the potential to establish norms.[122] In engaging with the *Plant Treaty* and the CBD the role and place of IP is undoubtedly in issue.[123]

Governing Body

The *Plant Treaty* established a Governing Body to promote the 'full implementation' of the *Plant Treaty*.[124] In implementing the IP obligations in a Material Transfer Agreement[125] the Governing Body negotiated a SMTA that addressed IP arrangements that were to apply to accessing the Multilateral System.[126] The SMTA is a contractual agreement between a Provider of materials[127] covered by the *Plant Treaty*'s Multilateral System[128] and a Recipient for use or conservation 'for the purposes of research, breeding and training for food and agriculture' (excluding 'chemical, pharmaceutical and/or other non-food/feed industrial uses').[129] In providing the materials the Provider undertakes that: 'Access to Plant Genetic Resources for Food and Agriculture protected by intellectual and other property rights shall be consistent with relevant international agreements, and with relevant national laws.'[130]

In return the Recipient undertakes: 'The Recipient shall not claim any [IP] or other rights that limit the facilitated access to the Material provided under this Agreement, or its genetic parts or components, in the form received from the Multilateral System.'[131] Also:

---

120 Ibid., Appendix B (Annex: [230]).

121 See CGRFA-12/09/Report, above n 38, [11] and Appendix G ([36], [39] and Indicative Timetable and Processes (p. 29)).

122 See ibid., [37] and [39].

123 This is already apparent in considering the specific issues raised by the *Plant Treaty* in revising the Global Plan of Action and its future iterations: see, for example, IT/GB-3/09/Report, above n 29, [50] and Appendix A.7 (Resolution 7/2009; [10]).

124 *International Treaty on Plant Genetic Resources for Food and Agriculture*, Art 19.3.

125 Ibid., Art 12.4.

126 IT/GB-1/06/REPORT, above n 28, [12] (Resolution 2/2006) and Appendix G (Standard Material Transfer Agreement).

127 Ibid., Appendix G (Standard Material Transfer Agreement) cl 5.

128 Ibid., Appendix G (Standard Material Transfer Agreement) cl 4.1.

129 Ibid., Appendix G (Standard Material Transfer Agreement) cl 6.1.

130 Ibid., Appendix G (Standard Material Transfer Agreement) cl 5(d).

131 Ibid., Appendix G (Standard Material Transfer Agreement) cl 6.2.

> A Recipient who obtains [IP] rights on any Products [being 'Plant Genetic Resources for Food and Agriculture that incorporate the Material or any of its genetic parts or components thereof that are ready for commercialization, excluding commodities and other products used for food, feed and processing'[132]] developed from the Material or its components, obtained from the Multilateral System, and assigns such [IP] rights to a third party, shall transfer the benefit-sharing obligations of this Agreement to that third party (footnote omitted).[133]

In short, the intention is that IP should not be claimed over materials received by the Recipient under an SMTA, albeit any changes or modifications by the Recipient may be subject to IP claims. Where materials are outside the scope of the *Plant Treaty* and to be used for purposes other than 'the purposes of research, breeding and training for food and agriculture' (excluding 'chemical, pharmaceutical and/or other non-food/feed industrial uses')[134] then IP can be claimed. This analysis also shows that the SMTA (and *Plant Treaty*) are limited in scope to 'facilitated access' to the materials in the Multilateral System *only* for use or conservation for the purpose of 'research, breeding and training for food and agriculture'.[135] Any other uses, including 'chemical, pharmaceutical and/or other non-food/feed industrial uses',[136] are outside the bounds of the SMTA and will require a separate agreement with no express obligations to benefit-share under the *Plant Treaty*. In short, the *Plant Treaty* and SMTA limit the scope to material provided for the purpose of 'research, breeding and training for food and agriculture'. Material provided for these other purposes might be provided on more or less favourable and different terms and conditions (albeit subject to the mandatory obligations of the CBD from 29 December 1993). The meaning of these terms 'the purposes of research, breeding and training for food and agriculture' (excluding 'chemical, pharmaceutical and/or other non-food/feed industrial uses') are unclear leaving considerable scope for access and benefit sharing outside the bounds of the *Plant Treaty* and its mandatory benefit sharing arrangements in the SMTA.

The CGIAR Centers

A central element to the *Plant Treaty* was the intention that it should include the germplasm collections of the CGIAR Centers. To this end the *Plant Treaty* expressly provided: 'The Multilateral System shall also include the plant genetic resources

132 Ibid., Appendix G (Standard Material Transfer Agreement) cl 2 ('Product').

133 Ibid., Appendix G (Standard Material Transfer Agreement) cl 6.10.

134 Ibid., Appendix G (Standard Material Transfer Agreement) cl 6.1.

135 Ibid., Appendix G (Standard Material Transfer Agreement) cl 6.1. See also *International Treaty on Plant Genetic Resources for Food and Agriculture*, Art 12.3(a).

136 Ibid., Appendix G (Standard Material Transfer Agreement) cl 6.1. See also *International Treaty on Plant Genetic Resources for Food and Agriculture*, Art 12.3(a).

for food and agriculture listed in Annex I and held in the *ex situ* collections of the International Agricultural Research Centers of the [CGIAR].'[137]

The *Plant Treaty* then makes provision for each of the CGIAR Centers to enter into agreements with the *Plant Treaty*'s Governing Body, making all the CGIAR Centers' *ex situ* collections available under the *Plant Treaty*.[138] Agreements have been entered into and the CGIAR Centers' collections are now part of the *Plant Treaty*'s Multilateral System accessed using the *Plant Treaty*'s SMTA.[139] Albeit these collections are within the *Plant Treaty*'s Multilateral System,[140] the CGIAR centers maintain an alternative evolving legal architecture. This legal architecture reflects the delicate balance between the CGIAR centers conducting their agricultural research activities and the funders providing the necessary funding for that research.

The CGIAR was established in 1971 under the auspices of the World Bank with funding from a range of sources including the FAO and the United Nations Development Program (UNDP).[141] As such, CGIAR started out as a funding arrangement for agricultural research targeting various existing institutions and programs that subsequently evolved into institutions that developed and maintained *ex situ* collections.[142] Through the decades these *ex situ* collections expanded to become the significant global resource.[143]

A key event in the evolution of these *ex situ* collections was the decision in 1994 in agreements signed with the FAO to bring the collections under the auspices of

137 *International Treaty on Plant Genetic Resources for Food and Agriculture*, Art 11.5.

138 Ibid., Art 15.1.

139 See Fourth Session of the Governing Body. 2011. *Report on the Implementation of the Multilateral System of Access and Benefit Sharing*, IT/GB-4/11/12. Rome: Governing Body, [35].

140 A notable exception are non-Annex 1 listed materials received and conserved by the CGIAR centers after the *Plant Treaty*'s entry into force (29 June 2004) that are made available on terms mutually agreed between the country of origin and the CGIAR Center according to the *Convention on Biological Diversity* (terms mutually agreed and prior informed consent): *International Treaty on Plant Genetic Resources for Food and Agriculture*, Art 15.3.

141 See, for an authoritative history of the creation of the CGIAR, Baum, above n 11, 3-59.

142 See ibid., 58-82.

143 Pistorious, above n 3, 48-68. See also Consultative Group on International Agricultural Research. 2011. *A Strategy and Results Framework for the CGIAR (approved by the Fund Council on 6 April 2011 and Consortium Board on 20 February 2011)*. Montpellier: Consultative Group on International Agricultural Research, [43] and Box 1.1.

the FAO.[144] In 2009 the CGIAR then undertook a major reform.[145] The outcome has been to clearly delineate the responsibilities and accountability between research and funding through formal adoption of the *Agreement Establishing the Consortium of International Agricultural Research Centers as an International Organization* (signed 13 September 2011). This agreement established the Consortium of International Agricultural Research Centers (CGIAR Consortium) as 'an independent international organization with full international legal personality'[146] and sets out its rules of operation in a constitution annexed to the Agreement.[147] The CGIAR Consortium presently consists of the 15 International Agricultural Research Centers (Member Centers or CGIAR Centers):[148]

---

144 Commission on Plant Genetic Resources. 1995. *Report of the Commission on Plant Genetic Resources*, CPGR-6/95/REP. Rome: Commission on Plant Genetic Resources, [14]. See also Commission on Plant Genetic Resources. 1995. *Joint report by FAO and the International Plant Genetic Resources Institute (on behalf of the CGIAR Centres) on the implementation of the agreement signed between FAO and the CGIAR Centres on 26 October 1994*, CPGR-6/95/12 Add.1. Rome: Commission on Plant Genetic Resources.

145 See Consultative Group on International Agricultural Research. 2010. *CGIAR Annual Report 2010: Towards a Reformed* CGIAR. Rome: CGIAR Secretariat, 1-8; Consultative Group on International Agricultural Research. 2009. *CGIAR Annual Report 2009: From Research to* Results. Rome: CGIAR Secretariat, 12-13; Consultative Group on International Agricultural Research. 2008. *CGIAR Annual Report 2008: Global Recommitment to Agriculture*. Rome: CGIAR Secretariat, 14. See also McAllister, E., Bezanson, K., Chadha, G., Mugabe, J. and Waage, J. 2008. *Bringing Together the Best of Science and the Best of Development: Independent Review of the CGIAR System – Report to the Executive Council*. Rome: CGIAR Secretariat.

146 *Agreement Establishing the Consortium of International Agricultural Research Centers as an International Organization*, Art 1. Notably this agreement is not presently in force albeit the CGIAR Consortium exists according to a contractual join venture signed on 29 April 2010.

147 Ibid., Art 3 and Annex (*Constitution of the Consortium of International Agricultural Research Centers*).

148 *Constitution of the Consortium of International Agricultural Research Centers*, Art 3. Being: Africa Rice Center (WARDA); Bioversity International (IPGRI); Centro Internacional de Agricultura Tropical (CIAT); Center for International Forestry Research (CIFOR); Centro Internacional de Mejoramiento de Maiz y Trigo (CIMMYT); Centro Internacional de la Papa (CIP); International Center for Agricultural Research in the Dry Areas (ICARDA); International Crops Research Institute for the Semi-Arid Tropics (ICRISAT); International Food Policy Research Institute (IFPRI); International Institute of Tropical Agriculture (IITA); International Livestock Research Institute (ILRI); International Rice Research Institute (IRRI); International Water Management Institute (IWMI); World Agroforestry Centre (ICRAF); Worldfish (ICLARM); and any new centers admitted by a three-quarters majority approval.

> The purpose of the Consortium shall be to provide leadership to the CGIAR system[149] and coordinate activities among Member Centers[150] and other Partners[151] within the framework of the *Strategy and Results Framework*[152] and the *CGIAR Research Programs*,[153] in order to enable them to enhance their individual and collective contribution to the achievement of the *CGIAR Vision*.[154]

The *CGIAR Vision* is: To reduce poverty and hunger, improve human health and nutrition, and enhance ecosystem resilience through high-quality international agricultural research, partnership and leadership.[155] The *CGIAR Vision* is to be implemented 'work[ing] with partners, stakeholders and potential beneficiaries towards three strategic objectives':

> (i) Create and accelerate sustainable increases in the productivity and production of healthy food by and for the poor (Food for People);
>
> (ii) Conserve, enhance and sustainably use natural resources, including biodiversity, to improve the livelihoods of the poor in response to climate change and other factors (Environments for People); and
>
> (iii) Promote policy and institutional change that will stimulate agricultural growth and equity to benefit the poor, especially rural women and other disadvantaged groups (Policies for People).[156]

The organs of the CGIAR Consortium are a Board and a Chief Executive Officer.[157] The Board (including the Chief Executive Officer and nine others selected and

---

149 'CGIAR system' meaning 'the global network of donors, other stakeholders, and international agricultural research Centers of the CGIAR, including all governing and advisory bodies': *Constitution*, n 148 above, Art 2.

150 'Member Centers' meaning 'the Member Centers of the Consortium': ibid.

151 'Partners' meaning 'Parties outside of the Consortium, Member Centers, Fund, donors, Trustee, and their affiliates, with which the Consortium and Member Centers will cooperate to further their collective purposes': ibid.

152 '*Strategy and Results Framework*' meaning 'the CGIAR Strategy and the framework to translate it into measurable developmental impact targets': ibid.

153 '*CGIAR Research Programs*' meaning 'research or other programs for coordinated implementation of the *Strategy and Results Framework*': ibid.

154 Ibid., Art 4.

155 *A Strategy and Results Framework*, above n 143, [48].

156 Ibid., [48] and footnote. See also CGIAR Working Group on Visioning. 2008. *Visioning the Future of the CGIAR, Report to the Executive Council 6 June 2008*. Rome: Consultative Group on International Agricultural Research, 14-15.

157 *Constitution of the Consortium of International Agricultural Research Centers*, Art 6.

appointed by the Member Centers)[158] provides strategy development, resource mobilization, funds allocation, set common policies and standards for CGIAR research programs, establishment of common reporting standards for performance and efficiency, and provide reports on its activities.[159] In addition to the Board and the Chief Executive Officer, the CGIAR Consortium Office (managed by the Chief Executive Officer)[160] assists the Board and Chief Executive Officer to undertake their work.[161] The CGIAR Consortium is also required to 'support the Member Centers in discharging their research and management roles', and Member Centers 'shall empower the CGIAR Consortium Board and CEO to execute their functions and powers' and 'shall accept the authority of the CGIAR Consortium Board and the CEO in decisions taken in exercise of these functions and powers'.[162] The effect of these arrangements is that the CGIAR Consortium Board (assisted by the Chief Executive Officer and the CGIAR Consortium Office) oversees implementation of the *Strategy and Results Framework*, decides on Member Centers' funding allocations and sets the policy applicable to the Member Centers. None of this is possible without funding – which is provided by donors through the CGIAR Fund (of which the Fund Council is the decision-making body) and through bilateral funding – and various donors have a direct involvement in settling the *Strategy and Results Framework* and the Member Centers' research programs (through the Funders' Forum).[163]

Reflecting the delicate balance between the CGIAR Centers conducting agricultural research and relying on donations for funding that research, the various donors have retained a considerable influence on the CGIAR Centers. This is reflected in the current governance arrangements with the Funders' Forum and Fund Council imposing various obligations on the CGIAR Consortium and consequently the CGIAR Centers. This is effected mostly through the *Joint Agreement* and the *Common Operational Framework* agreed between the Fund Council representing donors and the CGIAR Consortium.[164] In practice this means a CGIAR Research Program is proposed and then approved by the Fund Council, whereupon it is financed through contracts between the Fund Council and the CGIAR Consortium, then between the CGIAR Consortium and the lead Center, and then another contract between the lead Center and the Partner or other Center performing the contracted research activity. The terms and conditions of

158 Ibid., Art 7 and Exhibit 1.

159 Ibid., Art 8.

160 Ibid., Art 13(2).

161 Ibid., Arts 8(2)(e)(i) and 13(2).

162 Ibid., Art 15(2).

163 See ibid., Arts 5(2) and 8(2)(a).

164 See Consultative Group on International Agricultural Research. 2011. *Joint Agreement (approved by the Fund Council on 5 April 2011 and Consortium Board on 5 May 2011)*. Montpellier: Consultative Group on International Agricultural Research, [1.1]-[32.15] (*Joint Agreement*) and sch 3 (*CGIAR Principles*).

these contracts embody the various obligations set out in the various agreements between the Fund Council and the CGIAR Consortium including the *Strategy and Results Framework*, the *Common Operational Framework* (that incorporates significant policy positions including the *CGIAR Principles on the Management of Intellectual Assets*), and the *CGIAR Principles*, and so on. In limited circumstances donors can directly fund the Centers and specific projects conducted in Centers (known as 'Window 3' funds or bilateral funding).[165]

The *Joint Agreement* between the Fund Council and the CGIAR Consortium expressly deals with IP and provides in using funds:

> 9.1. The Fund Council requires that, in the context of [*Strategy and Results Framework*]-related activities, intellectual assets (which, for purposes of this *Joint Agreement*, shall include associated [IP] rights) be managed in a way that maximizes their global accessibility and ensures that they lead to the broadest possible impact on the poor, especially farmers in developing countries, consistent with the strategic objectives of the CGIAR.
>
> 9.2. For this purpose, the Consortium, in consultation with the Centers and Fund Council, shall develop a set of principles regarding the acquisition, management and release of intellectual assets within the context of [*Strategy and Results Framework*], which shall be included as a chapter in the *Common Operational Framework*, with the goal of submission of such a chapter for approval by the Fund Council and Consortium by September 30, 2011.
>
> 9.3. To the extent appropriate, each [CGIAR Research Program] proposal shall contain specific provisions regarding the management of the intellectual assets likely to result from the activities carried out under the [CGIAR Research Program], which shall be consistent with the set of principles referred to above, unless otherwise proposed by the Consortium and approved by the Fund Council.[166]

The *Strategy and Results Framework* in addressing IP provides:

> Although the production of public goods continues to be at the core of CGIAR's business, the legal boundaries for access and exchange of germplasm, technologies and research tools have changed considerably over the past few decades. [IP] rights have been progressively introduced in agriculture with the adoption of plant variety protection and patents over life forms and are increasingly accepted as the standard for the business and are no longer solely a private sector issue. International treaties – *International Union for the Protection of New Varieties of Plants* (UPOV), *International Treaty on Genetic Resources*

165 See ibid., sch 4 (*Form of Contribution Agreement*, Annex 2, B [1.1(iii)].

166 Ibid., [9].

> *for Food and Agriculture* (*Plant Treaty*), and the *Convention on Biological Diversity* (CDB) have also significantly evolved to provide a framework for the management of genetic resources and technologies. In this context, the CGIAR is developing clear system-wide IP principles taking into account the evolving situation with regards to the subject, as well as the realities and policies of the various types of partners with which it interacts. These principles will guide the management of intellectual assets, including the release of intellectual products that are produced by the Consortium and its member Centers as well as the acquisition of third party proprietary rights. These principles will be applied by the Consortium and its member centers across their activities and programs.[167]

In dealing with 'Intellectual Property'[168] and 'Intellectual Assets',[169] the *CGIAR Principles on the Management of Intellectual Assets*, as part of the *Common Operational Framework* and consistent with the *Joint Agreement*,[170] provide:

> All Intellectual Assets produced or acquired by the Consortium and/or the Centers shall be managed in ways that maximize their global accessibility and/or ensure that they lead to the broadest possible impact on target beneficiaries in furtherance of the *CGIAR Vision*.
>
> 6.1 **Prompt Dissemination of Research Results**.
>
> The Consortium and the Centers shall promptly and broadly disseminate their research results, subject to confidentiality obligations as may be associated with restrictions permitted under Articles 6.2 and 6.3 below, or subject to limited delays to seek [IP] under Article 6.4 below.

---

167 *A Strategy and Results Framework*, above n 143, [178].

168 'Intellectual property rights' means 'ownership rights (or applications for protection) of Intellectual Assets, whether registered or not, granted in any jurisdiction, including but not limited to, copyright and related rights, database rights, patents, industrial design rights, plant variety rights, trademarks and service marks, geographical indications, and trade secrets': Consultative Group on International Agricultural Research. 2011. *CGIAR Principles on the Management of Intellectual Assets (approved by the Fund Council on 7 March 2012 and Consortium Board on 1 March 2012)*, Montpellier: Consultative Group on International Agricultural Research footnote 5.

169 'Intellectual Assets' means 'any results or products of research and development activities of any nature whatsoever (including, but not limited to, knowledge, publications and other information products, databases, improved germplasm, technologies, inventions, know-how, processes, software, and distinctive signs), whether or not they are protected by [IP]': ibid., footnote 3.

170 See Consultative Group on International Agricultural Research. 2011. *Joint Agreement (approved by the Fund Council on 5 April 2011 and Consortium Board on 5 May 2011)*. Montpellier: Consultative Group on International Agricultural Research, [9].

6.2 **Limited Exclusivity Agreements**

6.2.1 The Consortium and/or the Centers may grant limited exclusivity for commercialization of the respective Intellectual Assets they produce ('Limited Exclusivity Agreements'), provided that:

(a) Such exclusivity is:

(i) Necessary for the further improvement of such Intellectual Assets or to enhance the scale or scope of impact on target beneficiaries, in furtherance of the *CGIAR Vision*; and

(ii) As limited as possible in duration, territory and/or field of use; and

(b) The Limited Exclusivity Agreements provide that the Intellectual Assets remain available[171] in all countries:

(i) For non-commercial research conducted by public sector organizations[172] in furtherance of the *CGIAR Vision* ('Research Exemption'); and

(ii) In the event of a national or regional Food Security Emergency[173] for the duration of the emergency ('Emergency Exemption').

6.2.2 Exemptions mentioned above, provided that this deviation is first approved by the Consortium. To obtain such approval, the Member Center shall submit a request in writing to the Consortium:

(i) Showing that the Limited Exclusivity Agreement meets the conditions provided under Article 6.2.1 (a);

---

171 'Available' means 'free-of-charge (except for actual costs or reasonable processing fees) or at a reasonable cost ["on the condition that this possibility of charging fees does not divert them from the fulfillment of the CGIAR Vision" except *Plant Treaty* materials to which the *Plant Treaty* will apply]': *CGIAR Principles on the Management of Intellectual Assets*, above n 168, Art 7 and footnote 6.

172 'Public sector organizations' means 'government entities, such as national governments, national agricultural research institutions, publicly funded international agriculture research centers, and publicly funded educational institutions': ibid., footnote 7.

173 'Food Security Emergency' means 'a food security related occurrènce that poses imminent threat of a significant loss of human life and which is declared an "emergency" by a national government or a multilateral and internationally recognized institution based on generally accepted benchmarks, such as the "level 4 emergency" or "level 5 catastrophe" categories of the Integrated Food Security Phase Classification (IPC)': ibid., footnote 8.

(ii) Explaining the reasons for the requested deviation; and

(iii) Showing that the deviation does not jeopardize the furtherance of the *CGIAR Vision*. The Consortium shall determine whether the reasons are compelling.[174] The Consortium shall have 30 days from notification to approve by affirmative consent or reject such request.

6.2.3 For the avoidance of doubt, the requirements set forth in article 6.2.1 above shall apply subject to third party restrictions permitted under Article 6.3 below.

6.3 **Incorporation of Third Party Intellectual Assets**

The Consortium and/or the Centers may enter into agreements for the acquisition and use of third party Intellectual Assets that restrict the global accessibility of the products/services resulting from the use of such Intellectual Assets for commercialization, research and development ('Restricted Use Agreements'), provided that:

(a) They are, to the best of their knowledge, unable to acquire equivalent Intellectual Assets from other sources under no or less restrictive conditions;

(b) The products/ services that are intended to result from the use of such third party Intellectual Assets will further the *CGIAR Vision* in the countries where they can be made available; and

(c) The Consortium and/or the Centers shall use their best efforts to ensure that such third party Intellectual Assets are only used in relation to, or incorporated into, such intended products/services.[175]

6.4 **Intellectual Property Rights**

6.4.1 The CGIAR is committed to the prudent and strategic use of [IP].

6.4.2 The Centers shall carefully consider whether to register/ apply for (or allow third parties to register/apply for) patents and/or plant variety protection ('IP Applications') over the Centers' respective Intellectual Assets. As a general principle, such IP Applications shall not be made unless they are necessary for

---

174 For the purposes of this decision 'the Consortium may take into account the source of funding, the potential for impact, the alternative means of access to the Intellectual Assets, the opportunity cost, etc.': ibid., footnote 9.

175 For example, 'Centers shall use best practices in field, greenhouse, laboratory, and livestock structure management to reduce the risk of unintended contamination of CGIAR biological materials by/with third party Intellectual Assets': ibid., footnote 10.

> the further improvement of such Intellectual Assets or to enhance the scale or scope of impact on target beneficiaries, in furtherance of the *CGIAR Vision.*
>
> 6.4.3 All [IP] shall be exercised consistently with Articles 6.1 to 6.3 above.[176]

Eleven of the CGIAR Centers have entered into agreements with the Governing Body of the *Plant Treaty*[177] and have adopted the *Plant Treaty*'s model SMTA subject to some minor modifications.[178] These modifications clarify that the SMTA may be used by the CGIAR Centers for materials held by the Centers but outside the scope of the Multilateral System (the transfers of non- Annex I materials).[179]

Global Crop Diversity Trust

The Global Crop Diversity Trust is an international fund with its own international legal personality originally established through agreement between the FAO and CGIAR:[180]

> ... it is the understanding of the Parties to this Agreement that the Governing Body of the [*Plant Treaty*] and the [Global Crop Diversity Trust] will enter into a separate agreement, recognizing the [Global Crop Diversity Trust] as an essential element of the Funding Strategy of the [*Plant Treaty*] and providing that the [Global Crop Diversity Trust] will operate under the overall policy guidance of the Governing Body of the [*Plant Treaty*].[181]

The objective of the Trust is 'to ensure the long-term conservation and availability of plant genetic resources for food and agriculture with a view to achieving global

176 Ibid., Art 6.

177 *CGIAR Principles on the Management of Intellectual Assets*, above n 168, Preamble. Notably early commitments were for all Centres to enter such agreements: IT/GB-1/06/REPORT, above n 28, [32] and Appendix K.

178 Governing Body of the International Treaty on Plant Genetic Resources for Food and Agriculture. 2007. *Report of the Second Session of the Governing Body of the International Treaty on Plant Genetic Resources for Food and Agriculture*, IT/GB-2/07/REPORT. Rome: Governing Body, [68].

179 Ibid., [68].

180 IT/GB-1/06/REPORT, above n 28, [38]-[39] and Appendix M (*Relationship Agreement between the Global Crop Diversity Trust and the Governing Body of the International Treaty on Plant Genetic Resources for Food and Agriculture*). See also Commission on Plant Genetic Resources. 1995. *Ninth Regular Session of the Commission on Genetic Resources for Food and Agriculture*, CGRFA-9/02/REP. Rome: Commission on Plant Genetic Resources, [49]-[50].

181 IT/GB-1/06/REPORT, above n 28, Appendix M (*Relationship Agreement between the Global Crop Diversity Trust and the Governing Body of the International Treaty on Plant Genetic Resources for Food and Agriculture*; Preamble).

food security and sustainable agriculture'.[182] The Governing Body (through the FAO) entered into this agreement,[183] and in setting a Funding Strategy as required by the *Plant Treaty*,[184] resolved that 'that the Global Crop Diversity Trust is an essential element of the Funding Strategy in relation to the *ex situ* conservation and availability of plant genetic resources for food and agriculture'.[185] The Governing Body subsequently

> recognized the executive independence of the [Global Crop Diversity Trust], and stressed the need for closer and effective cooperation. It emphasized that guidance for the [Global Crop Diversity Trust] was provided under Article 5, 6 and 17 of the *Global Plan of Action* and Article 5 and 17 of the [*Plant Treaty*].[186]

Article 5, 6 and 17 of the *Global Plan of Action* providing for sustaining existing *ex situ* collections, regenerating threatened *ex situ* accessions, and constructing comprehensive information systems for plant genetic resources for food and agriculture respectively.[187] And Article 5 and 17 of the *Plant Treaty* deal with the conservation, exploration, collection, characterization, evaluation, documentation and developing and strengthening information systems.[188] The Global Crop Diversity Trust has subsequently addressed these matters in its funding programs and decisions,[189] including funding *ex situ* collections for various crops in CGIAR Centers.[190]

---

182 Ibid., Appendix M (*Relationship Agreement between the Global Crop Diversity Trust and the Governing Body of the International Treaty on Plant Genetic Resources for Food and Agriculture*; Annex, Art 2(1)).

183 See ibid., [38]-[39].

184 *International Treaty on Plant Genetic Resources for Food and Agriculture*, Art 18.

185 IT/GB-1/06/REPORT, above n 28, [11] (Resolution 1/2006; Preamble, [(ix)]).

186 /GB-2/07/Report, above n 178, [59].

187 ITCPGR/96/REP, above n 24, [15] and Annex 2 (*Global Plan of Action for the Conservation and Sustainable Utilization of Plant Genetic Resources for Food and Agriculture*).

188 *International Treaty on Plant Genetic Resources for Food and Agriculture*, Arts 5 and 17.

189 See IT/GB-4/11/Report, above n 107, [34]; IT/GB-3/09/Report, above n 29, [34]. See also Governing Body of the International Treaty on Plant Genetic Resources for Food and Agriculture. 2011. *Report from the Global Crop Diversity Trust*, IT/GB-4/11/20. Rome: Governing Body, [8] and Appendix; Governing Body of the International Treaty on Plant Genetic Resources for Food and Agriculture. 2009. *Report from the Global Crop Diversity Trust*, IT/GB-3/09/10. Rome: Governing Body, [6], Annex and Appendix.

190 See, for example, Global Crop Diversity Trust. 2009. *Annual Report 2009*. Rome: Global Crop Diversity Trust, 6 and 13.

Of particular interest is the Global Crop Diversity Trust's support of the Svalbard Global Seed Vault.[191] The Global Crop Diversity Trust is a partner in the Svalbard Global Seed Vault with the Nordic Gene Bank and the Norwegian Ministry of Agriculture and Food,[192] and provides operational funding and targeted funding to assist CGIAR Centers to send their collections as duplicates to the vault.[193] Accessions are deposited with the Svalbard Global Seed Vault subject to an agreement between the 'Depositor' and the 'Royal Norwegian Ministry of Agriculture and Food'.[194] The agreement provides that:

a. The act of depositing seeds has 'no affect whatsoever on the nature and extent of any property rights pertaining to the Deposited Materials'.
b. 'The Depositor shall have the right to withdraw all or any of the Deposited Materials at any time on the giving of written notice'.
c. Where seeds have been deposited that are covered by the *Plant Treaty* (that is Annex 1 materials in the Multilateral System and non-Annex 1 materials held by the CGIAR Centers received before 29 June 2004) then the Depositor must make those seeds available to other natural or legal persons with the associated available non-confidential information according using the *Plant Treaty*'s standard material transfer agreement. CGIAR materials collected after 29 June 2004 and other non-Annex 1 materials must have been collected according to the *Convention on Biological Diversity*'s mutually agreed terms and prior informed consent, and these must be made available by the Depositor respecting the terms and conditions of collection.
d. May be terminated 'if the Depositor fails to comply fully with the terms and conditions set out in this Agreement' or 'for reasons of *force majeure*'.

In effect the Svalbard Global Seed Vault acts merely as a deposit and not as a point of access other than to the depositor. Where the Depositor ceases to exist or the agreement is terminated by reason of *force majeure* it is not clear what happens to the seeds or their ownership.

---

191 See, for example, Global Crop Diversity Trust. 2010. *Annual Report 2010*. Rome: Global Crop Diversity Trust, 6-7; *Annual Report 2009*, ibid., 13; Global Crop Diversity Trust. 2008. *Annual Report 2008*. Rome: Global Crop Diversity Trust, 14. See also IT/GB-4/11/20, above n 189, [23].

192 *Annual Report 2008*, above n 191, 14.

193 See *Annual Report 2009*, above n 190, 13; *Annual Report 2008*, above n 191.

194 See *Agreement Between Depositor and the Royal Norwegian Ministry of Agriculture and Food Concerning the Deposit of Seeds in the Svalbard Global Seed Vault*: http://www.nordgen.org/sgsv/scope/sgsv/files/SGSV_Deposit_Agreement.pdf. [Online]. (accessed: 1 February 2013).

World Intellectual Property Organization

The World Intellectual Property Organization (WIPO) was encouraged by the Conference of the Parties to the CBD to 'make rapid progress in the development of model IP clauses'.[195] This followed WIPO convening the Intergovernmental Committee on Intellectual Property and Genetic Resources, Traditional Knowledge and Folklore in 2001.[196] After protracted discussions about genetic resources, the work of the WIPO committee, 'without prejudice to the work pursued in other fora'[197] and considering the 'international dimension',[198] resolved a number of contentious matters:

a. The 'defensive protection of genetic resources' through the patent system[199] – These are measures to prevent the grant of a patent where there is a misappropriation (or misuse) of genetic resources ('biopiracy'), and to prevent wrongly or illegitimate granting patents because that do not meet the threshold requirements of novelty, inventive step, industrial applicability, and so on.[200] These discussions crystallized to a series of options: (i) an inventory of databases and information resources; (ii) various information systems about genetic resources for defensive protection; and (iii) various guidelines and recommendations on defensive protection.[201]

---

195 Conference of the Parties to the Convention on Biological Diversity. 2002. *Report of the Ad Hoc Open-Ended Working Group on Access and Benefit-Sharing*, UNEP/CBD/COP/6/6. Montreal: Secretariat of the Convention on Biological Diversity, 38 (Recommendations: Role of intellectual property rights in the implementation of access and benefit-sharing arrangements). See also UNEP/CBD/COP/6/20, above n 53, [56].

196 See Intergovernmental Committee on Intellectual Property and Genetic Resources, Traditional Knowledge and Folklore. 2001. *Report: First Session*, WIPO/GRTKF/IC/1/13. Geneva: World Intellectual Property Organization, [1].

197 World Intellectual Property Organization General Assembly. 2009. *Thirty-Eighth WIPO General Assembly: Report*, WO/GA/38/20. Geneva: World Intellectual Property Organization, [217]. See also World Intellectual Property Organization General Assembly. 2003. *Thirtieth WIPO General Assembly: Report*, WO/GA/30/8. Geneva: World Intellectual Property Organization, [93].

198 WO/GA/30/8, [93].

199 Intergovernmental Committee on Intellectual Property and Genetic Resources, Traditional Knowledge and Folklore. 2008. *Genetic Resources: List of Options,* WIPO/GRTKF/IC/13/8(a). Geneva: World Intellectual Property Organization Annex I([1]).

200 Intergovernmental Committee on Intellectual Property and Genetic Resources, Traditional Knowledge and Folklore. 2006. *The Patent System and Genetic Resources*, WIPO/GRTKF/IC/9/13. Geneva: World Intellectual Property Organization Annex ([24]-[36]).

201 Intergovernmental Committee on Intellectual Property and Genetic Resources, Traditional Knowledge and Folklore. 2010. *Genetic Resources: Revised List of Options*

b. The 'disclosure requirements in patent applications for information related to genetic resources used in the claimed invention'[202] – Related to the defensive measures, these are measures requiring patent and plant breeders' rights applications to disclose information about the genetic resources and traditional knowledge, such as evidence of prior informed consent, mutually agreed terms, and so on.[203] The discussions crystallized to a series of options: (i) mandatory disclosures; (ii) a further examination of issues relating to disclosure requirements; (iii) various guidelines and recommendations on defensive protection; and (iv) various alternative mechanisms.[204]
c. The 'IP issues in mutually agreed terms for the fair and equitable sharing of benefits arising from the use of genetic resources'[205] – These are measures that assume equitable benefits sharing from using genetic resources is through mutually agreed terms between the provider and user (bioprospector) according to a permit or a negotiated license, contract or agreement, and where IP is addressed as a term or condition of mutual agreement.[206] These have essentially been reduced to guidance about the kinds of terms and conditions that might be agreed between the provider and user (bioprospector) of genetic resources.[207] The discussions crystallized to a series of options: (i) an online database of IP clauses in mutually agreed

and Factual Update, WIPO/GRTKF/IC/17/6. Geneva: World Intellectual Property Organization, [5] and Annex 1([3]-[5]).

202 WIPO/GRTKF/IC/13/8(a), above n 199, Annex I([1]).

203 See, for example, Intergovernmental Committee on Intellectual Property and Genetic Resources, Traditional Knowledge and Folklore. 2007. *Declaration of the Source of Genetic Resources and Traditional Knowledge in Patent Applications: Proposals by Switzerland*, WIPO/GRTKF/IC/11/10. Geneva: World Intellectual Property Organization Annex; Intergovernmental Committee on Intellectual Property and Genetic Resources, Traditional Knowledge and Folklore. 2005. *Disclosure of Origin or Source of Genetic Resources and Associated Traditional Knowledge in Patent Applications*, WIPO/GRTKF/IC/8/11. Geneva: World Intellectual Property Organization Annex.

204 WIPO/GRTKF/IC/17/6, above n 201, [5] and Annex 1([6]-[11]).

205 WIPO/GRTKF/IC/13/8(a), above n 199, Annex I([1]).

206 See Intergovernmental Committee on Intellectual Property and Genetic Resources, Traditional Knowledge and Folklore. 2004. *Genetic Resources: Draft Intellectual Property Guidelines for Access and Equitable Benefit-Sharing*, WIPO/GRTKF/IC/7/9. Geneva: World Intellectual Property Organization. See also Intergovernmental Committee on Intellectual Property and Genetic Resources, Traditional Knowledge and Folklore. 2004. *Genetic Resources: Draft Intellectual Property Guidelines for Access and Benefit-sharing Contracts*, WIPO/GRTKF/IC/6/5. Geneva: World Intellectual Property Organization.

207 See WIPO/GRTKF/IC/7/9, above n 206, Annex.

terms about access and benefit-sharing; (ii) draft guidelines for contractual practices; and (iii) a study of licensing practices.[208]

Along the way WIPO has also developed *Draft Intellectual Property Guidelines for Access and Equitable Benefit-Sharing*,[209] an online database of biodiversity-related access and benefit-sharing agreements,[210] and a glossary of key terms related to IP and genetic resources.[211] In addition, and in response to a request from the Secretariat of the Conference of the Parties to the CBD to examine the 'interrelation of access to genetic resources and disclosure requirements in [IP] rights applications',[212] a 'draft' examination was prepared and considered,[213] and

208 WIPO/GRTKF/IC/17/6, above n 201, [5] and Annex 1([12]-[15]).

209 Intergovernmental Committee on Intellectual Property and Genetic Resources, Traditional Knowledge and Folklore. 2010. *Genetic Resources: Draft Intellectual Property Guidelines for Access and Equitable Benefit-Sharing: Updated Version*, WIPO/GRTKF/IC/17/INF/12. Geneva: World Intellectual Property Organization.

210 Intergovernmental Committee on Intellectual Property and Genetic Resources, Traditional Knowledge and Folklore. 2010. *Genetic Resources: Note on Updating of WIPO's Online Database of Biodiversity-Related Access and Benefit-Sharing Agreements*, WIPO/GRTKF/IC/17/INF/11, Geneva: World Intellectual Property Organization.

211 Intergovernmental Committee on Intellectual Property and Genetic Resources, Traditional Knowledge and Folklore. 2010. *Genetic Resources: Glossary of Key Terms Related to Intellectual Property and Genetic Resources*, WIPO/GRTKF/IC/17/INF/13. Geneva: World Intellectual Property Organization.

212 Conference of the Parties to the Convention on Biological Diversity. 2004. *Report of the Seventh Meeting of the Conference of the Parties to the Convention on Biological Diversity*, UNEP/CBD/COP/7/21. Montreal: Secretariat to the Convention on Biological Diversity, [347] and Annex (Decision VII/19; 281); Intergovernmental Committee on Intellectual Property and Genetic Resources, Traditional Knowledge and Folklore. 2004. *Genetic Resources: Certain Decisions of the Seventh Conference of the Parties to the Convention on Biological Diversity*, WIPO/GRTKF/IC/6/13. Geneva: World Intellectual Property Organization, [4].

213 See World Intellectual Property Organization General Assembly. 2005. *Thirty-Second WIPO General Assembly: Report*, WO/GA/32/13. Geneva: World Intellectual Property Organization, [211]; World Intellectual Property Organization General Assembly. 2005. *Invitation to WIPO from the Conference of Parties of the Convention on Biological Diversity*, WO/GA/32/8. Geneva: World Intellectual Property Organization Annex; Ad hoc Intergovernmental Meeting on Genetic Resources and Disclosure Requirements. 2005. *Compilation of Comments Received on the Second Draft of an Examination of Issues Relating to the Interrelation of Access to Genetic Resources and Disclosure Requirements in Intellectual Property Rights Applications Subsequent to an Ad Hoc Intergovernmental Meeting on Genetic Resources and Disclosure Requirements*, WIPO/IP/GR/05/INF/5. Geneva: World Intellectual Property Organization; Ad hoc Intergovernmental Meeting on Genetic Resources and Disclosure Requirements. 2005. *Examination of Issues Relating to the Interrelation of Access to Genetic Resources and Disclosure Requirements in Intellectual Property Rights Applications: Second Draft*, WIPO/IP/GR/05/3. Geneva: World Intellectual

submitted to the Conference of the Parties to the CBD.[214]

WIPO has continued its discussions with a renewed mandate and impetus for concluding the text of an agreement.[215] Despite this new mandate the negotiations remain protracted, albeit there exists a heavily bracketed consolidated text that continues to be negotiated.[216]

Conclusions

This chapter has addressed the sources of law making authority for IP for *ex situ* collections of plant germplasm. The outcome of this analysis in this chapter is to demonstrate that the existing international agreements establish some basic IP obligations. In implementing these obligations the various institutions are developing policy and guidance materials about how, where and when IP is to be addressed in negotiating access and benefit sharing. It is these policy and guidance materials that are directing practices that are establishing the norms for IP. At this stage it is too early to assess the actual impact of IP other than to note the potential of IP to affect both the price and availability of food, feed and fibre

---

Property Organization; Ad hoc Intergovernmental Meeting on Genetic Resources and Disclosure Requirements. 2005. *Compilation of Observations and Comments Submitted by Member States and Observers on the First Draft of an Examination of Issues Relating to the Interrelation of Access to Genetic Resources and Disclosure Requirements in Intellectual Property Rights Applications*, WIPO/IP/GR/05/INF/2. Geneva: World Intellectual Property Organization; Ad hoc Intergovernmental Meeting on Genetic Resources and Disclosure Requirements. 2005. *Examination of Issues Relating to the Interrelation of Access to Genetic Resources and Disclosure Requirements in Intellectual Property Rights Applications: First Draft*, WIPO/IP/GR/05/1. Geneva: World Intellectual Property Organization; Ad hoc Intergovernmental Meeting on Genetic Resources and Disclosure Requirements. 2005. *Member State Proposals and Suggestions Regarding the Invitation to WIPO from the Conference of the Parties to the Convention on Biological Diversity on Access to Genetic Resources and Disclosure Requirements in Intellectual Property Rights Applications*, WIPO/IP/GR/05/INF/1. Geneva: World Intellectual Property Organization.

214 Conference of the Parties to the Convention on Biological Diversity. 2006. *Interrelation of Access to Genetic Resources and Disclosure Requirements in Applications for Intellectual Property Rights: Report of the World Intellectual Property Organization* (WIPO), UNEP/CBD/COP/8/INF/7. Montreal: Secretariat to the Convention on Biological Diversity.

215 WIPO General Assembly. 2011. *Fortieth WIPO General Assembly: Report*, WO/GA/40/19. Geneva: World Intellectual Property Organization, [181]. See also WIPO General Assembly. 2011. *Matters Concerning the Intergovernmental Committee on Intellectual Property and Genetic Resources, Traditional Knowledge and Folklore (IGC)*, WO/GA/40/7. Geneva: World Intellectual Property Organization, [16] and [19].

216 See Intergovernmental Committee on Intellectual Property and Genetic Resources, Traditional Knowledge and Folklore. 2012. *Report: Twentieth Session*, WIPO/GRTKF/IC/20/10. Geneva: World Intellectual Property Organization, [714] and Annex II.

in the future. Whether this will be problematic is a question for the future. At this stage, however, it is apparent that there are various sites for norm-setting IP standards and that these norms will be about the ways in which IP is negotiated and practised. This is likely to be significant as in some circumstances (such as the kinds of public good research and exploitation practised by the CGIAR centers) is likely to develop a range of responses from no IP through the various extents of protection establishing IP through norm setting rather than just varying the existing threshold standards (such as those required by TRIPS).

Chapter 5

Open Access Seeds and Breeds: The Role of the Commons in Protecting Farmers' and Livestock Keepers' Rights and Food Security

Brendan Tobin

Introduction

> *Open access is now the default, and it is now the time to decide what we are going to do to make it happen.*
>
> Michael Marus – CGIAR Consortium Office (2012)

In a time of unparalleled wealth and food abundance there are still over 870 million people chronically undernourished in the world.[1] This food insecurity is due not so much to a lack of food as lack of access to food for the poorest sectors of society. Despite significant efforts in recent decades to reduce poverty and improve food production, food security continues to evade millions of people due to increased food costs caused by climate change, financial speculation, increased production of non-food cash crops, and the progressive consolidation of control over crop and breed diversity through the use of IP and other market tools. The grant of private property rights over lands, biological resources and the products of research and development using genetic resources and traditional knowledge poses particular threats to farming and livestock breeding communities dependent upon traditional rights of access to 'common property resources'. In response to the expansion of private property rights a dynamic alliance has emerged as indigenous peoples, farmers and livestock keepers, research institutions, international *ex situ*

1 Food and Agriculture Organization. 2012. *The State of Food Insecurity in the World 2012*. [Online]. Available at: http://www.fao.org/publications/sofi/en [accessed: 13 February 2013]. See also Jose, L.V. 2012. *A Binding Food Treaty: A Post-MDG Proposal Worth Exploring*. [Online]. Available at: http://www.falternativas.org/opex/documentos/memorandos/a-binding-food-treaty-a-post-mdg-proposal-worth-exploring [accessed: 13 February 2013].

collections, liberal and developing country states and international human rights bodies have bonded together to resist the continuing enclosure of the commons.

Building upon human rights law and development of *sui generis* legislative, contractual and quasi-contractual tools, this resistance is leading to the emergence of new constructed commons to secure local and global access to the land, resources and knowledge necessary to secure local and global food security and well-being. This chapter examines the potential role of an open access genetic resource commons for the protection of farmers and livestock keepers' rights and achievement of food security. The chapter begins by outlining the importance of agrobiodiversity to food security, as well as some of the threats posed by IP regimes and concentration of the seed and breed sectors for local and global food security. The chapter then explores a range of legal measures that have been proposed and/or adopted to protect indigenous and local community rights over their genetic resources and traditional knowledge, and the extent to which such measures help secure holistic protection of their resource and knowledge rights. In doing so the chapter discusses the limitations of existing international law for the protection of farmers' and livestock keepers' rights and the need for a more expansive approach that helps secure their land, resource and knowledge rights. It goes on to consider the potential merits of proposals for protection of farmers' and livestock keepers' rights through construction of a protected 'open access' genetic commons and the vital role that indigenous peoples and local communities' own legal regimes and decision-making authorities have to play in ensuring the protection of their resources, land and culture.

The chapter concludes by arguing that open access licensing regimes have an important role to play in securing a protected genetic commons and, thus, aiding the recovery of seed and breed sovereignty of local farmers and livestock keepers. This does not, however, mean that open access is the panacea of food security. The effectiveness of open access mechanisms will depend upon how they are integrated within the wider bundle of legal and quasi legal instruments and influences including human rights, broader IP regimes, farmers' and livestock breeders' rights, community legal initiatives, *sui generis* laws for protection of traditional knowledge, international compliance mechanisms (such as disclosure of origin in IP applications), and increased recognition and protection of indigenous peoples and local communities' customary laws and biocultural protocols. Considered as a complex, this bundle of rights has the capacity to secure the holistic and expansive protection of what the late Darrell Posey described as 'Traditional Resource Rights' – a unified rights concept that identifies the inseparable relationship between cultural and biological diversity, as well as human rights such as Indigenous and local communities' right to development and environmental conservation.[2]

2 Posey, D. 1996. *Traditional Resource Rights: International Instruments for Protection and Compensation for Indigenous Peoples and Local Communities*. Gland and Cambridge: IUCN, 15.

The Importance of Agrobiodiversity for Local and Global Food Security

Global food security and the livelihoods of hundreds of millions of people are inextricably dependent upon effective management and continuing access to plant and animal genetic diversity.[3] Local farming and livestock keeping systems play a pivotal role in global food security. Locally developed seed varieties, for instance, provide 60-90 per cent of seed planted in developing countries and are vital for local food security and for replenishment of *ex situ* collections.[4] Likewise, up to 50 per cent of the sheep, yak, camelid and goatherds breeds of those living in the most inhospitable and marginal dryland areas in Africa, Asia, Latin America and the Near East are locally adapted breeds.[5] Local seeds and breeds are, however, severely endangered by a range of climatic, economic, social and cultural threats.

One of the most significant threats to local seeds and breeds is an ever-increasing reliance on monocultures. This is brought about to a large extent by the increasing reliance on a few dominant breeds and the associated concentration of farming and grazing lands in the hands of national elites and foreign private and corporate interests.[6] In Sri Lanka, for instance, the number of rice varieties has decreased from 2,000 in 1959 to less than 100 in 2012 of which 75 per cent descend from a common stock, while in the United States 50 per cent of the wheat crop is represented by a mere nine varieties.[7] In terms of animal genetic resources, the Global Databank on Animal Genetic Resources classify 21per cent of animal breeds as being 'at risk' with a further 36 per cent classified as of unknown status.[8] Scherf estimates that out of 5,330 livestock mammals over 704 breeds are already

---

3 Commission on Genetic Resources for Food and Agriculture. 2007. *State of the World's Animal Genetic Resources for Food and Agriculture*. Rome: Food and Agriculture Organization, xxxv.

4 Almekinders, C. 2001. *Management of Crop Genetic Diversity at Community Level*. [Online]. Available at: http://www.hubrural.org/IMG/pdf/gtz_almekind.pdf [accessed: 13 February 2013].

5 Food and Agriculture Organization of the United Nations. 2009. *Livestock Keepers – Guardians of Biodiversity*, FAO Animal Production and Health Paper No. 167. Rome: Food and Agriculture Organization, 10; Info Resources Focus. 2007. *The Livestock Revolution: An Opportunity for Poor Farmers*, Focus No. 1/07, 3. [Online]. Available at: http://www.inforesources.ch/pdf/focus07_1_e.pdf [accessed: 13 February 2013]. Commission on Genetic Resources for Food and Agriculture. 2009. *The Use and Exchange of Animal Genetic Resources for Food and Agriculture*, Background Study Paper 43. Rome: Food and Agriculture Organization, 8.

6 Pearce F. 2012. *The Land Grabbers: The New Fight over Who Owns the Earth*. London: Transword Publishers.

7 Ceccarelli, S. 2012. Living seed – breeding as co-evolution, in *Seed Freedom – A Global Citizens Report*, edited by V. Shiva et al. New Delhi: Navdanya, 39.

8 Commission on Genetic Resources for Food and Agriculture, above n 5, 10.

extinct,[9] with more than 60 breeds becoming extinct in the first six years of the twenty-first century.[10] Coupled with this, marginalization of traditional livestock production systems and the increasing consolidation of rights to land, water and forage in the hands of well-off farmers, urban residents or the government, is further accelerating resource erosion and undermining both farmers' and livestock keepers' rights.[11] The rise of monocultures – that is, the narrowing genetic base and the rapid erosion of crop diversity – has significant implications for local and global food security. At the same time, the failure to internalize the costs of conservation of local seed varieties and animal breeds means poor farmers and livestock breeders, who are, in Brush's words, 'often poor and members of ethnic minorities',[12] are in essence subsidizing global crop conservation.[13]

The reliance on monocultures, and its impact on local and global food security, is a human rights issue. The right of all persons to adequate food and freedom from hunger is set out in Article 11 of the United Nations *International Covenant on Economic Social and Cultural Rights* (ICESCR). Article 11.1 of the ICESCR requires States to recognize 'the right of everyone to an adequate standard of living for himself and his family, including adequate food'. Article 11.2 recognizes that more immediate and urgent steps may be needed to ensure 'the fundamental right to freedom from hunger and malnutrition'. All states are obliged to secure these rights through their national legal system.[14] The fundamental right to food is further fortified by the provisions of common Article 1 of both the ICESCR and the United Nations *International Covenant on Civil and Political Rights* (ICCPR), which provides that 'no people may be deprived of their means of subsistence'.

---

9 Scherf, B.D. (editor). 2000. *World Watch List for Domestic Animal Diversity*, 3rd Edition. Rome: Food and Agriculture Organization, 59.

10 Commission on Genetic Resources for Food and Agriculture, above n 5, 10.

11 Info Resources Focus, above n 5, 8.

12 Brush, S. B. 1991. A farmer-based approach to conserving crop germplasm. *Economic Botany*, 45(2), 153-165, 155.

13 Virchow, D. 1999, Conservation of plant genetic resources for food and agriculture: Main actors and the costs to bear. *International Journal of Social Economics*, 26(7/8/9), 1144-1161.

14 According to the United Nations Committee on Economic, Social and Cultural Rights the states' obligations include to: *respect* existing access to food and not take measures preventing such access; *protect* the right to food including through the regulation of the activities of patent holders and plant breeders to prevent use of IP in a manner which would impede realization of farmers' rights to food; and *fulfil* the right to food by taking measures to support their access to and utilization of resources and means to ensure their livelihood, including food security: Committee on Economic, Social and Cultural Rights. 1999. *Substantive Issues Arising in the Implementation of the International Covenant on Economic, Social and Cultural Rights: General Comment 12*, E/C.12/1999/5. New York: United Nations, [15]. See also General Assembly. 2009. *Seed Policies and the Right to Food: Enhancing Agrobiodiversity and Encouraging Innovation*, A/64/170. New York: United Nations, 4.

Appearing, as it does, in both the ICESCR and ICCPR, this provision is both binding and enforceable. And, as of 2011 the right to food has been explicitly or implicitly protected under 56 national constitutions, in a further 51 countries the right is recognized due to the direct applicability of international treaties in their territories, while ten countries have adopted framework legislation on the human right to food or food security.[15] In India, for example, the Supreme Court has interpreted the right to life under the Indian Constitution to include the right to food.[16]

What is being done to ensure that there is sufficient and nutritious food? Agro-biotechnology, and its promises of high yielding plant varieties, is often wheeled out as the answer to achieving food security. Increased production of food through new technologies does not, however, necessarily equate with increased access to either food or technology. Diversion of agricultural production to non-food cash crops for international markets and massive land grabs for commercial agricultural purposes are just two of a range of agro-industrial practices that are having a detrimental impact on food security.[17] Added to this, over the past 30 years or so, a drop in government investment in research, developments in biotechnology and the expansion of IP protection over life forms has provided the incentives for multinational corporations aggressively to enter the field.[18] The result has been the rapid consolidation of the seed industry and the transformation of research products into private goods through the use of IP.[19] By 1999 three-

---

15 Knuth, L. and Vidar, M. 2011. *Constitutional and Legal Protection of the Right to Food Around the World*. Rome: Food and Agriculture Organisation, 32. At the national level Brazil has adopted the most far-reaching protection: see *National Food and Nutritional Security Framework Law No. 11,346 of 15 September 2006* establishing the National Food and Nutritional Security System (SISAN) to guarantee the human right to adequate food.

16 For discussion of *People's Union for Civil Liberties v. Union of India & Others (PUCL)* and related cases see: Birchfield, L. and Corsi, J. 2010. The right to life is the right to food: People's Union for Civil Liberties v. Union of India & Others. *Human Rights Brief*, 17(3), 15-18; Shah, M. B. 2006. The Indian Supreme Court acknowledges the right to food as a human right. *Agriculture & Rural Development*, 2, 24-26. [Online]. Available at: http://www.conectas.org/IBSA/RighttofoodIndianSC.pdf [acccessed: 13 February 2013].

17 See Graham, A., Aubry, S., Kunnemann, R. and Monsalve Suarez, S. 2010. *Land Grab Study*, 5. [Online]. Available at: http://www.fian.org/fileadmin/media/publications/2010_07_LandGrabStudy.pdf [accessed: 13 February 2013]; Daniel, S. with Mittal, A. 2009. *The Great Land Grab: Rush for World's Farmland Threatens Food Security for the Poor*, 4. [Online]. Available at: http://www.oaklandinstitute.org/sites/oaklandinstitute.org/files/LandGrab_final_web.pdf [accessed: 13 February 2013].

18 Van Overwalle, G. 2009. *Patents in Agricultural Biotechnology and the Right to Food*. [Online]. Available at: http://www.fundp.ac.be/droit/crid/propriete/G.VAN%20OVERWALLE%20scope%20of%20Patent.pdf [accessed: 13 February 2013].

19 ETC Group. 2008. *Who Owns Nature? Corporate Power and the Final Frontier in the Commodification of Life*, Communique No. 100. [Online]. Available at: http://www.etcgroup.org/sites/www.etcgroup.org/files/publication/707/01/etc_won_report_final_color.pdf [accessed: 14 February 2013]. According to the ETC Group, the top ten

quarters of patents on transformation technologies and genetic resources were controlled by only seven firms, the so-called 'Gene Giants'.[20]

The effect of the consolidation of seeds and breeds has been supported and amplified by the adoption of the *Agreement on Trade-related Aspects of Intellectual Property* (TRIPS), which requires all states to provide some form of protection for plant varieties. The implementation of TRIPS has raised tensions between intellectual property, human rights and food security. The Sub-Committee for the Promotion of Human Rights of the UN Commission for Human Rights, for instance, finds 'conflicts between the IP rights regime embodied in the TRIPS Agreement, on the one hand, and international human rights law, on the other'.[21] The High Commissioner for Human Rights has criticized TRIPS, arguing that '[a] human rights approach ... would explicitly place the promotion and protection of human rights, in particular those in ICESCR, at the heart of the objectives of [IP] protection'.[22] The specific threat IP poses to farmers has been signalled by the UK government-appointed IPR Commission, which claims that '[IP] protection ... [carries] ... the risk of restricting farmers' rights to reuse, exchange and sell seed ... practices which form the basis of their traditional role in conservation and development'.[23] With regards to impacts on food security, De Schutter warns of the potential negative impacts of IP on accessibility to the means of food production.[24] Although obligations under the ICCPR and IESCR fall primarily on states, the Commission on Economic, Social and Cultural Rights highlighting the relationship between nation states, private business, and food security, states, in General Comment 12, that '[a]s part of their obligations to protect people's resource base for food, States parties should take appropriate steps to ensure that

---

seed companies account for 67 per cent of the global proprietary seed market; the world's largest seed company alone, Monsanto, accounts for 23 per cent of that market; and the top three companies (Monsanto, DuPont and Syngenta) account for 47 per cent of the market, including 65 per cent of the maize seed market and over half of the proprietary soybean seed market. It is this concentration, rather than the introduction of IP rights alone, that may result in excessively high prices for agricultural inputs.

20 Hope, J. 2008. *Biobazaar: The Open Source Revolution and Biotechnology*. Cambridge, MA: Harvard University Press, 65.

21 Commission on Human Rights, Sub-Commission on the Promotion and Protection of Human Rights. 2000. *Intellectual Property Rights and Human Rights*, Resolution 2000/7, E/CN.4/Sub.2/RES/2000/7. New York: United Nations.

22 United Nations. 2001. *The Impact of the Agreement on Trade Related Aspects of Intellectual Property Rights on Human Rights: Report of the High Commissioner*, E/CN.4/Sub.2/2001/13. New York: United Nations, [22].

23 Commission on Intellectual Property Rights. 2002. *Integrating Intellectual Property Rights and Development Policy: Report of the Commission on Intellectual Property Rights*. London: Commission on Intellectual Property Rights, 68.

24 A/64/170, above n 14, 5.

activities of the private business sector and civil society are in conformity with the right to food'.[25]

The tension between IP and food security is further complicated by international regulation of access to genetic resources and biological diversity. At the same time that IP regimes have expanded both geographically and in their scope, the global treatment of genetic resources has swung from one of largely unregulated access to one of jealously guarded sovereign rights, as recognized under the *Convention on Biological Diversity* (CBD). The result is a twofold squeeze on the commons. On the one hand, many states have assumed the right to regulate access to all genetic resources, including those belonging to farmers and livestock keepers. On the other hand, IP protection is further diminishing the scope and accessibility of previously freely available resources. For Kloppenburg the solution is clear, if true 'food sovereignty' is to be achieved then, he says, 'control over genetic resources must be wrested from the corporations and governments that seek to monopolize them and be restored to, and permanently vested in, social groups and/or institutions with the mandate to sustain them and to facilitate their equitable use'.[26] With the adoption of the *International Treaty on Plant Genetic Resources for Food and Agriculture* (*Plant Treaty*) the international community has made an initial step towards the construction of a virtual commons around a small group of genetic resources for food and agriculture. At the same time it has given token recognition to the notion of farmers' rights, and opened a door of opportunity that may yet have an important role to play in the struggle for recognition and protection of their rights to land, resources, knowledge and self-determination. How, then, are farmers' and livestock keepers' rights protected?

Legal Protection of Farmers' and Livestock Breeders' Rights

During the years immediately following the adoption of the TRIPS agreement it appeared as if nothing could halt the seemingly inexorable expansion of IP regimes and the associated corporate enclosure of the commons. As developing countries became resigned to the imposition of TRIPS, indigenous peoples, local communities and farmers' groups, around the world mobilized to oppose patents and other proprietary ownership rights over living things and their constituent parts. Incensed by iconic cases of biopiracy indigenous peoples took to the streets, courts and international fora, to defend their cultural and biological heritage.[27] This included challenges to notorious cases of biopiracy such as the patenting of

---

25 E/C.12/1999/5, above n 14, [27].

26 Kloppenburg, J. 2010. Seed sovereignty: The promise of open source biology, in *Food Sovereignty: Reconnecting Food Nature and Community*, edited by A. Aurélie Desmarais et al. Black Point: Fernwood Publishing Company Limited, 152.

27 Robinson, D.F. 2010. *Confronting Biopiracy: Challenges, Cases and International Debates* London: Earthscan, 51-53.

Ayahuasca, a sacred plant used for centuries by Amazonian peoples, and patents derived from use of neem, tumeric, maca, yellow beans and basmati rice. In many cases these patents have been shown to be fatally flawed leading Silvia Ribeiro to describe the IP system as 'broken beyond repair'.[28]

Denouncing biopiracy as another form of colonialism – colonial expropriation of traditional knowledge – played well among developing countries still smarting from the imposition of TRIPS. Framing the debate in this way also garnered support among liberal European states, international organizations with mandates in the area of development and human rights, research institutions concerned to show commitments to fair and equitable benefit sharing, and a host of international, local and issue based NGOs. The informal alliance that coalesced around traditional knowledge provided the basis for the first serious challenge to the progression of the IP. Such resistance alone would have been futile, however, were it not supported by proposals for legislative and other measures that could turn the tables on the powers supporting uncontrolled expansion of IP and in the process protect the rights of local communities and indigenous peoples over their genetic and cultural heritage. To be effective these proposals had to address not only IP issues but also issues of land, resource, knowledge and cultural rights, and their status under indigenous peoples and local communities' own legal regimes. Most importantly, such proposals had to ensure that the very process of protection of resources and knowledge did not undermine the cultural integrity of its custodians.

One of the earliest and most influential proposals was Posey's promotion of the concept of 'Traditional Resource Rights'.[29] In Posey's view, the effective protection of indigenous peoples and local communities' rights over their resources and knowledge required recognition and enforcement of a 'bundle of rights' including, most importantly, rights of self-determination[30] and:

> basic human rights, right to development, rights to environmental integrity, religious freedom, land and territorial rights, right to privacy, prior informed consent and full disclosure, farmers' rights, intellectual property rights, neighbouring rights, cultural property rights, cultural heritage recognition, rights of customary law and practice.[31]

---

28 ETC Group. 2009. *Enola Patent Ruled Invalid: Haven't We Bean here Before? (Yes, Yes, Yes, Yes and Yes.)*. [Online]. Available at: http://www.etcgroup.org/sites/www.etcgroup.org/files/publication/pdf_file/etcnr_enola_14july09.pdf [accessed: 14 February 2013].

29 Posey, above n 2.

30 Ibid., 17.

31 Posey D.A. 1995. *Indigenous Peoples and Traditional Resource Rights: A Basis for Equitable Relationships?* Workshop on Indigenous Peoples and Traditional Resource Rights. [Online]. Available at: http://www.ubcic.bc.ca/files/PDF/Posey_Indigenous.pdf [accessed: 14 February 2013].

There have been numerous other proposals dealing with indigenous peoples and local communities' knowledge and resource rights. Nijar, for example, took a more direct path than Posey by drafting a proposal for a community intellectual rights regime that eschewed individualistic proprietary rights in favour of a system of collective community control over resources.[32] Nijar's proposal was influential in the adoption of model legislation for Africa, which prohibits the patenting of life forms.[33] At the national and regional level legislation to protect biodiversity[34] and traditional knowledge[35] provides varying levels of recognition and protection to farmers belonging to indigenous peoples and local communities. In more recent times, proposals have emerged for the establishment of agrobiodiversity conservation zones as a means to protect local farmers' varieties.[36] At a more local level, indigenous peoples and local communities, often working in conjunction with NGOs, have established community seed banks, databases and registers of genetic resources and traditional knowledge.[37] Going even further, community conservation and biocultural heritage areas, such as the Potato Park in Peru, provide support for local conservation, repatriation and control over local seed varieties.[38]

32 *Nijar, G.S. 1996. In Defence of Local Community Knowledge and Biodiversity: A Conceptual Framework and the Essential Elements of a Rights Regime*. Penang: Third World Network, 56.

33 See Organization of African Unity. 2000. *African Model Legislation for the Protection of the Rights of Local Communities, Farmers and Breeders, and for the Regulation of Access to Biological Resources*. [Online]. Available at: http://www.cbd.int/doc/measures/abs/msr-abs-oau-en.pdf [accessed: 14 February 2013].

34 See, for example, Peru's Law No. 26839, 8 July 1997: *Law on the Conservation and Sustainable use of Biological Diversity* that declares the knowledge, innovations, and practices of indigenous communities are part of their cultural patrimony. See also Costa Rica *Biodiversity Law* No. 7788, 1998.

35 See, for example, Peru's Law No. 27811, 24 July 2002: *Regime for the Protection of Indigenous Peoples' Collective Knowledge Associated with Biodiversity*, Brazil's *Provisional Measure* No. 2.052 of 21 December 2000, Philippines' *Act to Recognize, Protect and Promote the Rights of Indigenous Cultural Communities/Indigenous Peoples* (No. 837, of 28 July, 1997).

36 Tapia, M. and Tobin, B. 2013. Guardians of the seed: Caring and sharing of agrobiodiversity in the Andes, in *A Common Pools Approach in International Biodiversity Law: Enhancing Sustainable Use of Genetic Resources*, edited by E. Kamau and G. Winter. London: Routledge, 77-79.

37 For discussion of a range of experiences with databases and registers see Alexander, M., Chamundeeswari, K., Kambu, A., Ruiz, M. and Tobin, B. 2004. *The Role of Registers and Databases in the Protection of Traditional Knowledge: A Comparative Analysis*. [Online]. Available at: http://www.ias.unu.edu/binaries/UNUIAS_TKRegistersReport.pdf [accessed: 14 February 2013].

38 Argumedo A. and Pimbert, M. 2006. *Protecting Indigenous Knowledge Against Biopiracy in the Andes*. [Online]. Available at: http://pubs.iied.org/pdfs/14531IIED.pdf? [accessed: 14 February 2013].

Another significant outcome of indigenous peoples' efforts to exercise their sovereign rights over their lands, resources and knowledge has been the development of biocultural protocols. The term 'biocultural protocols' is used to embrace a wide range of contractual agreements, codes of conduct and procedures for application regarding access to and use of their lands, resources and knowledge.[39] Forged by experience and drawing heavily upon their own laws, customs and practices, biocultural protocols have become an important tool for community empowerment and construction of interfaces between indigenous peoples' and local communities' legal regimes and state and international law.

Traditionally, IP is seen as antithetical to protecting the interests of indigenous peoples. In fact, indigenous peoples and local communities have strongly and consistently opposed the use of patents and other forms of IP to secure their ancestral rights over genetic resources and traditional knowledge. Barsh, for example, claims that a property rights approach will lead to 'a distortion of the very nature of indigenous cultures, and of the relationship between Indigenous peoples and their lands'.[40] This does not mean, however, indigenous peoples have completely rejected the possibility that some forms of IP may provide means for protecting their resource and knowledge rights. Trademarks, geographical indicators and copyright have all been looked to as potential means to secure rights over indigenous products and traditional cultural expressions. Their limited use to date demonstrates, however, their restricted utility. For instance, as the Federal Court opined in the case of *Yumbulul v Reserve Bank of Australia*, 'Australia's copyright law does not provide adequate recognition of Aboriginal community claims to regulate the reproduction and use of works which are essentially communal in origin'.[41] In the view of aboriginal peoples, the limitations of IP law make it insufficient to regulate rights over stories and rock paintings made thousands of years ago, and the onus is on national and international law to change to respect indigenous peoples' customary laws.[42]

39 For discussion and information on biocultural protocols visit the International Institute for Environment and Development. *Biocultural Heritage*. [Online]. Available at: http://biocultural.iied.org/tools/community-biocultural-protocols [accessed: 14 February 2013]; Community Protocols. *Community Protocols*. [Online]. Available at: http://www.community-protocols.org [accessed: 14 February 2013]. See also Kabir, B. and Jonas, H. 2009. *Biocultural Community Protocols: A Community Approach to Ensuring the Integrity of Environmental Law and Policy*. [Online]. Available at: http://naturaljustice.org/wp-content/uploads/pdf/BCPs_community_approach2009.pdf [accessed: 14 February 2013].

40 Barsh R.L. 1999. How do you patent a landscape? The perils of dichotomizing cultural and intellectual property. *International Journal of Cultural Property*, 8(1), 14-47, 15.

41 *Re Terry Yumbulul v Reserve Bank of Australia; Aboriginal Artists Agency Limited and Anthony Wallis* [1991] FCA 332, [21].

42 See statement of the Association of Northern and Kimberly Artists of Australia in Janke, T. 1999. *Our Culture: Our Future – Report on Australian Indigenous Cultural and Intellectual Property Rights*. Canberra: Michael Frankel & Co., 59.

One IP-based proposal that has gained widespread support[43] is for the modification of IP application procedures to require disclosure of the origin of genetic resources and traditional knowledge and evidence of prior informed consent as a condition for processing applications for patents and other IP.[44] The idea behind the disclosure of origin concept is to shift the burden of proof regarding rights to make use of genetic resources and traditional knowledge from its custodians to its users.[45] The disclosure concept has been adopted in part by Decision 6/24 of the CBD and the Nagoya Protocol, adopted in October 2010, provides for a international system of certification of compliance with national access and benefit sharing legislation, seen as a key element for a functional disclosure regime.[46] Developing countries, led by Brazil, Peru and India, have made numerous proposals for modification of TRIPS at the WTO where a majority of member states now, reportedly, support some form of binding disclosure requirements.[47] Disclosure requirements are now found in the legislation of many developed and developing countries[48] and have also been incorporated into regional law by the Andean community.[49] Disclosure issues are also a key element of current negotiations at WIPO for the development of new international instruments on genetic resources, traditional knowledge and traditional cultural expressions.

---

43 Tobin, B. 1994. *Alternativas a Las Legislaciones de Propiedad Intelectual.* [Online]. Available at: http://www.academia.edu/1315002/Alternativas_a_la_Propiedad_Intelectual [accessed: 14 February 2013].

44 By focusing on the application procedure rather than the conditions for grant of patents or other IP the proposal avoids a conflict with TRIPS and other IP instruments. No challenge has been brought against any national or disclosure of origin regime under the WTO. See Sarnoff, J. and Correa, C. 2005. *Analysis of Options for Implementing Disclosure of Origin Requirements in IP Applications.* [Online]. Available at: http://unctad.org/en/Docs/ditcted200514_en.pdf [accessed: 14 February 2013].

45 Tobin, B. 1997. *Certificates of Origin: A Role for IPR Regimes in Securing Prior Informed Consent.* [Online]. Available at: http://www.ias.unu.edu/sub_page.aspx?catID=5&ddlID=370 [accessed: 19 February 2013].

46 Tobin, B., Burton, G. and Fernandez-Ugalde, J. C. 2008. *Certificates of Clarity or Confusion: The Search for a Practical, Feasible and Cost Effective System for Certifying Compliance with PIC and MAT.* [Online]. Available at: http://www.ias.unu.edu/sub_page.aspx?catID=7&ddlID=682 [accessed: 14 February 2013].

47 Kohr, M. 2008. *TRIPS: Majority of WTO Members now Support Disclosure Proposal.* [Online]. Available at: http://www.twnside.org.sg/title2/wto.info/twninfo20080318.htm [accessed: 14 February 2013].

48 Amongst the countries that have adopted some form of disclosure regimes are Norway, Belgium, Germany, Egypt, Brazil, Bolivia, Ecuador, Peru, Colombia, Costa Rica, India. The Andean Community is the first regional Economic Block to adopt a disclosure of origin regime.

49 Andean Community. *Decision 486: Common Intellectual Property Regime*, Article 26(i). [Online]. Available at: http://www.sice.oas.org/trade/junac/decisiones/dec486e.asp [accessed: 14 February 2013].

All the foregoing initiatives have in their own way helped to advance national and global protection of indigenous peoples' and local communities' rights to their resources and knowledge. While it is still too early to determine the impact of existing mechanisms for protection of their rights, by slow, sometimes faltering and piecemeal steps, Posey's concept of a holistic system of 'Traditional Resource Rights' is taking shape. A key component of that vision is the concept of farmers' rights, an issue to which we now turn.

Rights of Farmers and Livestock Keepers

The notion of farmers' rights, first proposed as a foil for plant breeders' rights, has been given legal recognition in the *Plant Treaty*.[50] Although widely decried as being too-little, too-late and too-vague, the very recognition of a right in favour of farmers may prove significant in the longer term. This is the view of livestock keepers', for instance, who are calling for the formalization of farmers' rights through an international legal regime that recognizes and protects their rights and interests. Just how these rights have evolved and the extent to which they are recognized is the subject of the following analysis.

Farmers' Rights

Since the 1930s, the adoption of laws permitting the patenting of plants and the granting of plant breeders rights has led to a progressive curtailing of farmers' freedom to save and sell their seed. At the outset plant breeders' rights as conceived by the 1961 *International Convention for the Protection of New Plant Varieties* (that set up the International Union for the Protection of New Plant Varieties (UPOV)) provided a farmers' privilege to collect and grow the seeds of their harvest. In UPOV 1991 the farmers' privilege as a right was replaced with a discretionary power exercisable by states.[51] As the farmers' privilege waned, Patrick Mooney and Cary Fowler 'concocted' a new concept of farmers' rights, which they successfully floated during negotiation of the FAO *International Undertaking on Plant Genetic Resources for Food and Agriculture* (IU). This led

50 Mooney P. 2011. International non-governmental organizations: The hundred year (or so) seed war – seeds, sovereignty and civil society – a historical perspective of the evolution of the 'law of the seed', in *Plant Genetic Resources and Food Security: Stakeholder Perspectives on the International Treaty on Plant Genetic Resources for Food and Agriculture*, edited by C. Frison et al. London: Earthscan, 135-148, 147.

51 Biber-Klemm, S. 2011. *Livestock Keepers' Rights: Draft for Discussion at the Workshop on Sui-generis Rights for Animal Genetic Resources for Food and Agriculture; Berne, World Trade Institute, 20 June 2011*, Working Paper No 2011/68. [Online]. Available at: http://www.nccr-trade.org/fileadmin/user_upload/nccr-trade.ch/wp3/3.5/Working%20 paper%202011%2068%20-%20Pdf.pdf [accessed: 14 February 2013].

to the inclusion of farmers' rights in the IU.[52] The term 'Farmers' Rights' was later defined in FAO Conference Resolution 5/89, as meaning 'rights arising from the past, present and future contributions of farmers in conserving, improving, and making available plant genetic resources, particularly those in the centres of origin/diversity'.

This language was later reflected in the Preamble to the *Plant Treaty*[53] that replaced the IU upon entering into force in 2004. The Preamble describes a range of rights 'recognized in the Treaty' as fundamental to the realization of farmers' rights including 'rights to save, use, exchange and sell farm-saved seed, and other propagating material, and to participate in decision-making, and in the fair and equitable sharing of the benefits arising from, the use of plant genetic resources for food and agriculture'. The substantive provisions of the *Plant Treaty*, however, leave the issue of state recognition of the farmers' privilege to use farm-saved seed 'entirely to national decision-makers'.[54] On a more positive note, Article 9.2 of the *Plant Treaty* goes beyond the preambular language including reference to traditional knowledge.[55] That said, the substantive provisions of the *Plant Treaty* have been prefaced by a chapeau that, according to Mooney, has 'rendered the strong affirmative language beneath almost irrelevant'.[56] The relevant text of Article 9.2 reads as follows:

> The Contracting Parties agree that the responsibility for realizing Farmers' Rights, as they relate to plant genetic resources for food and agriculture, rests with national governments. In accordance with their needs and priorities, each contracting Party should, as appropriate, and subject to its national legislation, take measures to protect and promote Farmers' Rights, including:
>
> (a) protection of traditional knowledge relevant to plant genetic resources for food and agriculture …;
>
> (b) The right to equitably participate in sharing benefits from the utilization of plant genetic resources for food and agriculture; and
>
> (c) The right to participate in making decisions, at the national level, on matters related to the conservation and sustainable use of plant genetic resources for food and agriculture.

---

52 Food and Agriculture Organization of the United Nations. 1989. *Report of the Conference of FAO*, Twenty-fifth Session. Rome: Food and Agriculture Organization, [108] (Resolution 5/89).

53 Moore G. and Tymoswski, W. 2005. *Explanatory Guide to the International Treaty on Plant Genetic Resources for Food and Agriculture*. Gland: IUCN, 67.

54 Ibid., 25.

55 Ibid.

56 Mooney, above n 50, 147.

Via Campesina, an international movement that brings together marginalized groups such as peasants, small- and medium-size farmers and women farmers, sees the text of the *Plant Treaty* as a backward step for farmers' rights, and a 'bleak lip service to what these rights should entail'.[57] Via Campesina has rejected any approach to implementation of farmers' rights that relies on distribution of the profits arising from exploitation of IP rights, calling instead for a total ban on IP over seeds and public policies in favour of 'living farmers' seed systems, 'systems that are in our communities and under our control'.[58] Far from securing farmers' collective rights over their local varieties, the concept of farmers' rights has, in Fowler's view, been 'hi-jacked' as a slogan for 'advocating a form of [IP] rights for farmers varieties' while failing to secure farmers' rights to conserve these resources.[59] In an attempt to salvage something from the debris, Argumedo et al., turn to other provisions of the *Plant Treaty* including Article 5.1 (in-situ conservation), and Article 6 (the requirements for law and policy to support diverse farming systems), which they say, 'call[s] for a broad interpretation of farmers' rights which goes beyond the right to benefit-sharing, to include the right of farmers to continue the practices which contribute to the conservation and sustainable use of PGRFA, and to sustain the traditional knowledge and livelihood systems needed for this'.[60]

A more expansive approach to farmers' rights is also required to meet obligations relating to access to genetic resources and protection of traditional knowledge under the CBD and the Nagoya Protocol. The Protocol requires all states to ensure that access to and use of traditional knowledge and genetic resources of indigenous and local communities is subject to the communities' prior informed consent. It also obliges states to take into consideration 'customary laws and protocols' in the implementation of the Protocol (Article 12). The obligations under the Nagoya Protocol are of significant importance for the realization of farmers' rights covering as they do issues of 'prior informed consent', distribution of benefits and access to justice. Furthermore, viewed from this perspective the *Plant Treaty* provides the legal basis to interpret farmers' rights as encompassing a wide range of rights including rights to food security, land, resources, knowledge, culture, human dignity and self-determination.

---

57 Svanhild-Isabelle Batta, B. 2004. *Breakthrough for 'The South'? An Analysis of the Recognition of Farmers' Rights in the International Treaty on Plant Genetic Resources for Food and Agriculture*. Lysaker: Fridtjof Nansen Institute, 67.

58 See First Meeting Ad Hoc Technical Advisory Committee on Sustainable Use of Plant Genetic Resources for Food and Agriculture, *Compilation of Views and Experiences on the Implementation of Farmers' Rights submitted by Contracting Parties and Relevant Organizations*, IT/ACSU-1/12/Inf.3. Rome: Governing Body, 39.

59 Svanhild-Isabelle Batta, above n 57, 94.

60 Argumedo, A., Swiderska, K., Pimbert, M., Song Y. and Pant, R. 2011. *Implementing Farmers' Rights under the FAO International Treaty on PGRFA: The Need for a Broad Approach Based on Biocultural Heritage*, 2. [Online]. Available at: http://pubs.iied.org/pdfs/G03077.pdf [accessed: 14 February 2013].

Furthermore, the relevant provisions of the *Plant Treaty*, the Nagoya Protocol and the CBD must be read in light of broader human rights law. Indeed, these rights in so far as they apply to indigenous peoples have already been recognized under international human rights instruments, such as *International Labour Organization Convention 169* (ILO Convention 169) and the United Nations *Declaration on the Rights of Indigenous Peoples* (UNDRIP). Taken together, these instruments recognize extensive rights in favour of indigenous and tribal peoples over their traditional lands, resources, knowledge and self-determination.[61] Furthermore, upwards of 100 national constitutions oblige national governments, in varying degrees, to give due respect and recognition to indigenous and tribal peoples' laws and traditions.[62]

Effective implementation of the Nagoya Protocol and broader human rights laws (that recognize the rights of indigenous peoples and communities) would empower indigenous peoples and local farming communities to control access to and use of their resources and knowledge in accordance with their own customary laws and practices. It would also ensure their rights to continue their traditional resource, management and sharing practices. However, the lack of a functional international mechanism for ensuring compliance with human rights law and the Nagoya Protocol threatens to undermine realization of these rights. Furthermore, early proposals for national and regional implementation of the Nagoya Protocol have failed to show the respect and recognition for customary law called for by the Protocol and human rights instruments.[63] As a consequence, ensuring the protection of farmers' rights in accordance with their own laws and practices will now fall to other fora, most particularly the World Intellectual Property Organization (WIPO). At WIPO on going negotiations on genetic resources, traditional knowledge and traditional cultural expressions provide a clear opportunity to recognize the role of customary law in issues such as the identification of rights over resources and knowledge, definition of prior informed consent procedures, adjudication of compliance and resolution of disputes.[64] Taken together with the Nagoya Protocol

61 See generally, Tobin, B. 2012. 'Why Customary Law Matters: The Role of Customary Law in the Protection of the Human Rights of Indigenous Peoples', PhD thesis Irish Centre for Human Rights, National University of Ireland Galway; Charters, C. and Stavenhagen, R. 2009. *Making the Declaration Work: The United Nations Declaration on the Rights of Indigenous Peoples*. Copenhagen: International Work Group For Indigenous Affairs.

62 Cuskelly, K. 2011. *Customs and Constitutions: State Recognition of Customary Law Around the World*. Bangkok: International Union for the Conservation of Nature, 4; Tobin, above n 61, 98 *et seq*.

63 Neither the process nor the content of proposals for national and regional legislation to implement the Nagoya Protocol developed by Australia, the European Union and Switzerland in the latter half of 2012 show that consideration has been given to indigenous peoples' customary laws and protocols in their development.

64 World Intellectual Property Organization. 2006. *Customary Law and the Intellectual Property System in the Protection of Traditional Cultural Expressions and*

and human rights instruments, the WIPO negotiations will play an important role in securing farmers' rights, and providing the recognition of their property interests in a manner that reflects their own notions of stewardship and collective ownership.

Despite the desultory nature of the international treatment of farmers' rights, some commentators have noted positive outcomes at a national level. Andersen and Winge, for example, highlight the Norwegian government's rejection of UPOV 1991 and of stricter plant breeders' rights in favour of continuing recognition of the farmers' privilege.[65] They also laud India's *Protection of Plant Varieties and Farmers' Rights Act* of 2001, which recognized farmers' rights to save, use, sow, re-sow, exchange, share and sell farm produce, including seeds of varieties protected by plant breeders' rights.[66] Such apparent advances may themselves, however, prove illusory. The Indian law, for example, has been criticized by the International Institute for Environment and Development (IIED), which claims that due to 'the vague system for registration and benefit sharing in the law and the inability of farmers to apply for registration, it seems extremely far-fetched that farmers' rights are going to be protected through the law'.[67] That said, it seems clear that responsible and judicious government action coupled with due respect for human rights and relevant multilateral environmental agreements has the power to direct implementation of farmers' rights towards securing-long term conservation of local varieties and biocultural heritage. What more is required? Political will, effective participatory processes for legislative development and program implementation and solid support from the international community.

Livestock Keepers' Rights

Livestock is a valuable asset for poverty reduction, playing a role in 'food security, nutrition, community health, socio-cultural identity, empowerment and gender

---

Knowledge, WIPO Issues Paper, Version 3.0. [Online]. Available at: http://www.wipo.int/export/sites/www/tk/en/consultations/customary_law/issues-revised.pdf [accessed: 14 February 2013]. See also Taubman, A. 2005. Saving the village: Conserving jurisprudential diversity in the international protection of traditional knowledge, in *International Public Goods and Transfer of Technology Under a Globalized Intellectual Property Regime*, edited by K.E. Maskus and J.H. Reichmann. Cambridge: Cambridge University Press, 521-564.

65 Andersen, R. and Winge, T. 2008. *Success Stories from the Realization of Farmers' Rights Related to Plant Genetic Resources for Food and Agriculture*, The Farmers' Rights Project Background Study 7, 15. [Online]. Available at: http://www.fni.no/doc&pdf/FNI-R0408.pdf [accessed: 14 February 2013].

66 Ibid., 13.

67 See submission made by the International Institute for Environment and Development: Governing Body of the International Treaty on Plant Genetic Resources for Food and Agriculture. 2010. *Compilation of Views and Experiences of the Implementation of Farmers' Rights Submitted By Contracting Parties and Relevant Organizations*, IT/GB-4/11/Inf.6. Rome: Governing Body, 15.

balance, natural resource management, risk management, access to services and markets, and as a source of income'.[68] While the notion of farmers' rights has been around since the early 1990s, the concept of livestock keepers' rights is of more recent vintage, being developed in a participatory process leading up to the Interlaken Technical conference on Animal Genetic Resources in 2007.[69] Although, there is no agreed definition of the term 'livestock keepers', there is a widespread tendency to distinguish them from specialized breeders such as private companies, cooperatives and state run breeding farms.[70] For the purpose of this study, 'livestock keepers' embrace the almost 70 per cent of the world's poor that keep livestock, including 100 million landless people, 640 million subsistence farmers in rain-fed areas and 190 million pastoralists in arid or mountainous areas.[71] Livestock keepers are also inextricably linked to indigenous peoples' and local communities' traditional knowledge systems.[72]

Livestock keepers, especially small-scale poor herders, play a vital role in conservation of animal genetic resource (AnGR) diversity. Unfortunately, however, there are a number of threats to their way of life, including their reliance on access to common resources, including pastures, forests and water.[73] Somewhat ironically, the onset of what has been termed a livestock revolution, which it is estimated will use 80 per cent of the agricultural land surface by 2020, is likely to weaken rather than strengthen the position of livestock keepers in general.[74] Increased production and competition for land threatens to squeeze poor livestock keepers even further onto marginal lands, while both land grabs and conservation policies bar them from their traditional grazing grounds. They also face threats from climate change, desertification and exposure to diseases, as well as global markets that poorly remunerate producers. Factors that, Rota and Sidahmed say, make 'finding a balance between food security and safety, poverty alleviation and sustainable use of natural resources a tremendous challenge'.[75] A further challenge for small-scale poor herders is the role IP is likely to play in market consolidation and restriction of access to technologies necessary to respond to climatic, health and market challenges ahead. Although patenting based on local breeds is not

---

68 Rota, A. and Sidahmed, A. 2010. *IFAD's Livestock Position Paper: Livestock Planning, Challenges and Strategies for Livestock Development in IFAD*, 8-9. [Online]. Available at: http://www.ifad.org/lrkm/factsheet/livestockpaper.pdf [accessed: 14 February 2013].

69 Biber-Klemm, above n 51.

70 Commission on Genetic Resources for Food and Agriculture, above n 5, 9.

71 Geerlings, E, E. Mathias and I Köhler-Rollefson. 2002. *Securing Tomorrow's Food: Promoting the Sustainable Use of Farm Animal Genetic Resources: Issues and Options*. [Online]. Available at: http://www2.gtz.de/dokumente/bib/02-5195.pdf [accessed: 14 February 2013].

72 Ibid., 9.

73 Rota and Sidahmed, above n 68, 10.

74 Ibid., 13.

75 Ibid., 16.

currently a problem, it is only a matter of time until patents are sought over modified breeds incorporating resistance traits from native breeds.[76]

In the absence of international legal recognition, the definition of livestock keepers' rights is found in instruments such as the *Declaration of Livestock Keepers' Rights*, adopted by livestock keepers meeting in South Africa in 2008.[77] The Declaration reflects the contrast between the seed trade and livestock transfers, the latter being primarily north–south, which establishes a distinct dynamic from that surrounding the struggle for farmers' rights. The lack of significant commercial interest in local breeds diminishes the influence of livestock keepers over international governance of AnGR. As a consequence, their demands as set out in the Declaration focus primarily on continued maintenance of their breeds and of their access to traditional grazing sites rather than on commercial benefits. The Declaration sets out three key principles and five rights. These are:

> Principles:
>
> Livestock Keepers are creators of breeds and custodians of animal genetic resources for food and agriculture.
>
> Livestock Keepers and the sustainable use of traditional breeds are dependent on the conservation of their respective ecosystems.
>
> Traditional breeds represent collective property, products of indigenous knowledge and cultural expression of Livestock Keepers.
>
> Livestock Keepers have the right to:
>
> make breeding decisions and breed the breeds they maintain.
>
> participate in policy formulation and implementation processes on animal genetic resources for food and agriculture.
>
> appropriate training and capacity building and equal access to relevant services enabling and supporting them to raise livestock and to better process and market their products.

76 Geerlings et al., above n 71, 77 where the authors report examples of corporate interest in among other things trypano tolerance of N'dama cattle and worm resistance of Red Maasai sheep.

77 *Declaration on Livestock Keepers Rights*. [Online]. Available at: http://ebookbrowse.com/declaration-on-lkrs-with-initial-signatories-6-pdf-d21842186 [accessed: 19 February 2013].

participate in the identification of research needs and research design with respect to their genetic resources, as is mandated by the principle of Prior Informed Consent.

effectively access information on issues related to their local breeds and livestock diversity.[78]

Although, as a group, livestock keepers are not recognized under international law, they made their presence known at the twelfth meeting of the CBD Conference of the Parties. There is still, however, no firm commitment by the international community to recognizing their rights despite significant reasons for doing so, not least the dangers of continuing erosion of livestock diversity and related threats to food security. As Kohler-Rollefson points out, locally adapted breeds are able to survive on marginal lands that are not suitable for crop cultivation and are the building blocks for future livestock development.[79] They secure livelihoods and food security in some of the most inhospitable environments, and in so doing they continue to breed resistance to harsh climatic conditions, pests and disease that may well prove vital for securing long-term global food security. An international treaty on animal genetic resources and livestock keepers' rights is indeed, therefore, warranted. The weakness of farmers' rights under the *Plant Treaty* and the lack of international recognition of livestock keepers' rights means it is necessary to look elsewhere for mechanisms to protect the rights of breeders and livestock keepers. One area of interest is the free and open access movement, whose earliest proponents were, Feingold claims, farmers themselves.[80]

Open Access as a Tool for Self-Empowerment

In the 1970s and early 1980s the free software movement emerged with the aim of promoting the development of computer operating systems in a decentralized non-proprietary fashion.[81] In 1984, Richard Stallman launched the GNU Project as a means for counteracting the growing domination of information technology

78 Ibid.

79 Kohler-Rollefston, I. 2005. *Building an International Legal Framework on Animal Genetic Resources: Can it Help the Drylands and Food-insecure Countries?* [Online]. Available at: http://www.pastoralpeoples.org/docs/int_legal_framework_an_gen_res.pdf [accessed: 14 February 2013].

80 Feingol, S. 2012. *The Hard Path to Open Source Bioinnovation: How Cambia is Strengthening the Agricultural Open Source Infrastructure, Science Progress.* [Online]. Available at: http://scienceprogress.org/2012/08/the-hard-path-to-open-source-bioinnovation [accessed: 14 February 2013].

81 Santilli, J. 2012. *Agrobiodiversity and the Law: Regualting Genetic Resources, Food Security and Cultural Diversity*. New York: Earthscan.

development by a small group of monolithic companies.[82] A key part of Stallman's approach was the development of a general licence that offered access to anyone who agreed to be bound by the terms of the licence. In 1985, the Free Software Foundation was set up to promote the 'broader development and use of free software'.[83] Twenty-five years later the free access software revolution is so successful that it has become the dominant provider of software for Internet servers.[84]

The open access software movement is not without its critics. In the 1990s use of the term 'free access' began to be seen as an obstacle to 'emerging business processes', which though openly accessible were not without cost.[85] This was largely a problem of terminology and the debate is, perhaps, best summed up by the Free Software Foundation who stated that the term 'free' means 'free as in "free speech", not as in "free beer"'.[86] As a consequence of the ambiguity over the term 'free', the preferred term now is 'open source'.[87] Successful open source technologies include Apache (with over 57 per cent of the Internet server market)[88] and the Linux operating system (without which Amazon would not be profitable[89] and upon which the Google Chrome operating system is based).[90] Open source is not, therefore, some offbeat reactionary movement, it is an integral part of our lives. Its success has inspired initiatives such as the open source journal the Public Library of Science (PLoS), to the Tropical Disease Initiative, which seeks to produce medications for 'orphan diseases' such as malaria, cholera, dengue fever and Chagas disease.[91] Open source is the kernel upon which a more

82 Hope, above n 20, 8.

83 Ibid., 9.

84 For an in-depth discussion of the open source movement see ibid.

85 Oldham, P.D. 2009. *An Access and Benefit-Sharing Commons? The Role of Commons/Open Source Licences in the International Regime on Access to Genetic Resources and Benefit-Sharing*, 4. [Online]. Available at: http://papers.ssrn.com/sol3/papers.cfm?abstract_id=1438027 [accessed: 14 February 2013].

86 See GNU Operating System. *The Free Software Definition.* [Online]. Available at: http://www.gnu.org/philosophy/free-sw.html [accessed: 14 February 2013].

87 As of February 2013 the Open Source Initiative (OSI) website listed 67 approved open source licences: www. http://opensource.org/licenses/alphabetical.

88 Derived from the data available at the Netcraft website www.netcraft.com (November 2012): Netcraft. *Market Share for Top Servers Across All Domains.* [Online]. Available at: http://news.netcraft.com [accessed: 14 February 2013].

89 Hope, above n 20, 295.

90 Oldham, above n 85, 5.

91 See Santilli, above n 81, 259. The Tropical Disease Initiative, a joint project of the Goldman School of Public Policy of the University of California Berkley, the Department of Biopharmaceutical Sciences and Pharmaceutical Chemistry of the University of California in San Francisco and Duke University Law School, has the 'aims to provide a "kernel" for open source drug discovery. Such kernel should allow scientists from laboratories, universities, institutes, and corporations to work together for a common cause: find new

communitarian humanistic approach to science and technology development and knowledge exchange is based. It is a kernel which some believe provides the basis for securing the rights of indigenous peoples and local communities over their resources and knowledge in a manner that reflects their traditional resource and knowledge management systems, while at the same time maintaining a vibrant resource and knowledge commons for the protection of global health and food security.

Kloppenburg suggests that a system of open source biology can help to recreate the commons by 'simultaneously pursuing *both* effective resistance *and* the creation of a protected space into which practices and institutions with truly transformative capacity can be introduced and elaborated'.[92] While open access is not, in his view, a panacea, he sees it as a plausible way of moving forward in the struggle for seed sovereignty.[93] Kloppenburg's view was informed by a proposal by bean breeder Tom Michaels who took Stallman's general licence and applied it to genetic resources, arguing that it could help protect public goods and prevent appropriation of the results of publicly funded research.[94] Kloppenburg, who describes Michael's proposed general public licence for genetic resource (GPLPGR) as 'simple, elegant and effective',[95] identifies four key benefits of such a GPLGR regime. These include: prevention of biopiracy; re-establishing farmers' seed sovereignty: providing the basis for increased collaboration between farmers, scientists and research institutions in participatory plant breeding programs; and helping to foster the exchange, sale and marketing of seed that is not covered by proprietary rights.[96] According to Kloppenburg, the benefits of open access regimes will lead to a more expansive notion of seed sovereignty involving a wide range of actors from farmers to scientists and CGIAR.[97]

While the theory of open source sounds convincing, in practice things may prove more complicated. To paraphrase Chander and Sunder, adopting and adapting their terminology, undue 'romance of the commons' fails to recognize the disparate levels of capacity and opportunity for use of free access resources and knowledge.[98] As they put it: 'the notion that when a resource is open to all by the force of law, all will be equally able to exploit it – obscures the harsh realities of a world fraught with inequality'.[99] Commercial and scientific exploitation of

---

drugs against tropical diseases such as Malaria or Tuberculosis': Tropical Disease Initiative. *In Brief.* [Online]. Available at: http://sgt.cnag.cat/TDI/ [accessed: 14 February 2013].

92 Kloppenburg, above n 26, 153.

93 Ibid., 157.

94 Ibid., 160.

95 Ibid., 161.

96 Ibid., 163.

97 Ibid.

98 Chander, A. and Sunder, M. 2004. Romance of the public domain. *California Law Review*, 92(5), 1331-1373.

99 Ibid., 1354.

traditional knowledge and genetic resources in an open access commons is likely to be dominated by developed countries, or more specifically by corporations based in developed countries. A study prepared by the Harvard Business School Life Sciences Project on the use of public biodatabases showed that more than 92 per cent of access was from users from 10 countries over half of whom were in the United States, and that over 50 per cent of that use was by corporate users. In the words of Martinez:

> a new world hierarchy is developing, one that separates those nations and regions that are bioliterate from those that are bio-illiterate. This is the world of DNA space, populated by a self-selecting few who have chosen to participate in the new technology revolution. The price of admission: the ability to produce, read, or translate DNA. This means that even as biodata begins to drive industries from agribusiness to computing, cosmetics to chemical manufacturing, few nations have the skills required to develop, access, and use it.[100]

Another drawback of open access regimes is a lack of clarity over who can put resources and knowledge into the open source system. This is an issue because resources and knowledge are widely shared within and among communities and peoples. In the context of this chapter, questions also arise over the extent to which farmers' and livestock keepers' rights over their local seeds and breeds may be restricted by national claims of sovereign rights over genetic resources. The open access concept underlying open source systems may also find itself at odds with the more restricted knowledge sharing commons in which indigenous resources and traditional knowledge circulate. This raises questions about what restrictions may be placed on access, and, ultimately the use of resources and knowledge in an open access system. Would restrictions on who can access resources, only men or only women, for example, be acceptable if they are necessary to protect the integrity of biocultural heritage? As one colleague framed this dilemma the pertinent question is: how open must open access be to be considered open for a global commons?[101]

Another challenge associated with open access regimes is their reliance on IP regimes for their protection. Initially, the success of the open source software movement in computer programs and software was facilitated by copyright law, which provided legal protection of the author's rights over their source code. As Oldham puts it: open source 'navigates the space between the public domain and proprietary models based on copyright and software patents'.[102] This creates

---

100 Martinez, R., Enriquez, J. and West, J. 2003. The geography of the genome. *Wired.* [Online, June]. Available at: http://www.wired.com/wired/archive/11.06/dna_spc.html [accessed: 14 February 2013].

101 Pers. comm., Dr Leanne Wiseman, ACIPA, Griffith Law School, December 2012.

102 Oldham, above n 85, 5.

ambiguity. The open source system does not exist outside of IP regimes but operates from within the system, seeking to use IP regimes to secure community rights while eschewing their enclosure in exclusive proprietary rights. This poses a number of significant problems for breeders and livestock keepers. Firstly, it requires indigenous peoples and local communities to embrace IP regimes as defenders of their rights: an idea that has largely been dismissed by the global indigenous movement. Secondly, as Hardison points out, indigenous peoples' and local communities' resources and knowledge navigate within their own resource and knowledge sharing spaces, subject to their own collective stewardship and bounded by their own legal regimes and practices.[103] Moreover, the establishment of new open access regimes may threaten the integrity of such local sharing spaces and the cultural reality they support. Thirdly, copyright, which, as we have seen, underpinned the open source software movement will not be as relevant in the context of breeders and livestock keepers. Rather, according to Kloppenburg, 'extant property rights' will provide the necessary protection for open access genetic resources. [104] Though he does not specify the source of such rights, it is reasonable to assume that Kloppenburg is referring to rights enshrined in indigenous peoples' and local communities' own legal regimes, in constitutional and national law, as well as, in international human rights and customary international law. More specifically, international legal instruments such as the Nagoya Protocol, CBD, *Plant Treaty*, UNDRIP and ILO Convention 169, as well as emerging legal instruments such as those under negotiation at the WIPO Inter-governmental Committee, provide clear support for the recognition of the rights of indigenous peoples and local communities to control access to and use of their resources in accordance with their own laws, traditions and customs. These laws, traditions and customs must be taken into consideration by all states in implementing the Nagoya Protocol.

When thinking about the issues involved in open access regimes one thing is clear: it is not going to be easy to develop a truly global protected commons. Convincing indigenous peoples to place their resources and knowledge in a global commons may prove more difficult, especially when they are 'so imbued with spiritual or cultural meaning that, even if they can be shared, it may be unacceptable to relinquish control over subsequent uses'.[105] What is the way forward? One proposal would be for a web of local and resource specific protected commons using a menu of licensing regimes allowing for varying degrees of control to be retained over conditions of access and use. There is no reason why distinct groups of farmers, livestock keepers, local communities or indigenous peoples should

103 Raven, M. 2005. Rethinking the public domain: A challenge for knowledge-sharing societies in the information age. *Work in Progress*, 17(2), 22-23.

104 Kloppenburg, J. 2010. Impeding dispossession, enabling repossession: Biological open source and the recovery of seed sovereignty. *Journal of Agrarian Change*, 10(3), 367-388.

105 Kloppenburg, above note 26, 328.

not construct their own bounded regimes, appropriating as practical elements of open access regimes further to consolidate their stewardship over their resources and knowledge. Based on research surrounding the development of biocultural protocols with various traditional communities, Abrell et al. propose the use of what they term a TK Commons licence to regulate a range of local commons such as a 'pastoralists commons' a 'traditional healers commons' and a 'Biosphere commons'.[106] Such a diverse regime might draw upon the experience of the legal models developed by the creative commons and scientific commons to develop a range of licensing instruments that may be modified to respond to the legal norms of the relevant indigenous people.[107]

Any scheme must also give thought to the strengths and limitations of licensing. Licences will, for example, only create a contractual obligation between the original parties to the contract and those sublicensees who agree to be bound by the terms of the original agreement. It does not bind any third party who might obtain access to the relevant resources or knowledge outside the framework of the licence. It is, therefore, merely a question of time before a breach of farmers' and livestock keepers' rights over their resources finds its way into the courts of a foreign jurisdiction. In such cases the existence of community biocultural protocols, some form of open access licensing regime and/or the customary laws of the farmers and livestock keepers may be crucial to proving the existence of their rights and sustaining an action arising from their breach.

In addition, licenses will only be enforceable to the extent that they conform to the law, which may vary greatly from country to country and even from state to state. While the parties are free to agree the terms and conditions they deem appropriate, if those terms and conditions conflict with national law, including IP law, a licence may be deemed unenforceable in whole or in part. Despite such weaknesses, licences are an integral part of a wider array of legal tools including indigenous peoples' own customary laws, tribal statutes, biocultural and other protocols. Because of this, licenses can help strengthen and where necessary re-establish their resource and knowledge commons safe from the threats of dispossession. If this is achieved it will in turn help to secure the long term viability of their seed and breed stock, its progressive development and their access to improved seeds and breeds free of cost. Such free access to improved seeds and breeds is one of the keys to ensuring local and global food security in the long term.

---

106 Abrell, E. et al. 2009. *Implementing a Traditional Knowledge Commons: Opportunities and Challenges*, 28. [Online]. Available at: http://naturaljustice.org/wp-content/uploads/pdf/Implementing_a_TKC-2009.pdf [accessed: 14 February 2013].

107 Oldham, above n 85.

Conclusion

The response of the international community to the erosion of crop and livestock diversity and to corporate consolidation in the seeds and breeds sectors has been both slow and inconsistent. On the one hand, the CBD and Nagoya Protocol have established a bilateral trading system governing access to genetic resources and benefit sharing that is unwieldy. On the other hand, the *Plant Treaty* creates a commons for 64 crop varieties and a weak system of farmers' rights that fails to capture an equitable share of commercial benefits and to internalize the costs of conservation of agrobiodiversity. The disparate approaches of the Nagoya Protocol and *Plant Treaty* to the governance of genetic resources have failed to secure the confidence of a sector vital for the conservation of genetic diversity. In addition, small farmers and livestock keepers (in particular those from indigenous and local communities), who are responsible for the maintenance of the vast majority of crop and breed diversity, are largely left without a voice. Despite their weaknesses, the *Plant Treaty*, CBD and the Nagoya Protocol complement, and are in turn complemented by, a wide range of other legislative, contractual and community-based measures that together are leading towards the realization of 'Traditional Resource Rights' as envisaged by Posey.

Open access offers a way for indigenous peoples and local communities to protect their resource and knowledge rights while maintaining or re-establishing local, regional or, even, global commons. Open access must, however, be seen as part of a wider set of instruments that together create a legal framework that establishes the boundaries of indigenous peoples' and local communities' rights over their resources and knowledge.[108] This includes their own legal regimes, customs and protocols; national biodiversity, genetic resource and traditional knowledge laws; constitutional recognition of their rights to self-determination and to their own laws and decision-making authorities; international human rights and environmental law; and, compliance mechanisms such as disclosure requirements in IP regimes and international systems for certification of compliance with their customary laws and protocols and national law. Incentives created by this framework will serve to promote compliance by those seeking access to and use of genetic resources and traditional knowledge. It is the existence of such incentives and compliance mechanisms that make an open access licensing regime over indigenous peoples' and local communities' resources and knowledge viable from a legal basis. It does not, however, resolve the key challenge of how to ensure that an open access licensing regime does not distort cultural and spiritual relationships to the land, resources and knowledge.

An open access regime is not a panacea, or a one-size-fits-all solution. It does, however, have an important role to play in conserving native breeds and seeds, for the benefit of both present and future generations and for both local and global

108 In some cases legal recognition of indigenous peoples' rights may be greater than those of local communities belonging to national minorities.

food security. In conjunction with the wide range of instruments identified in this chapter, open access licensing can serve as a tool for the empowerment of indigenous and local communities. It can help in the establishment of a functional bridge between their own customary and statutory legal regimes and national and international law and between their respective decision-making authorities. If this was all it achieved it would be a great result in itself, signifying a further healing in the breach between international and national positive law and the legal regimes of indigenous peoples: a breach that runs counter to human rights. Every step towards healing that breach brings the protection of indigenous peoples' and local communities' rights over their resources and knowledge closer. An open access regime can also play an important role in helping to redraw notions of property rights in a manner that respects indigenous peoples' notions of stewardship[109] based on concepts of responsibility and reciprocity over their resources and knowledge. In so doing it can help promote increased intercultural legal pluralism[110] and equity.[111] Together these various steps can help secure food sovereignty, security and the protection of diversity. Without them, the future for crop and breed diversity is bleak.

---

109 Carpenter, K.A., Katyal, S. and Riley, A. 2009. In defense of property. *Yale Law Journal*, 118(6), 101-204, 145.

110 For discussion of the notion of intercultural legal pluralism see Tsosie, R. 2007. Cultural challenges to biotechnology: Native American genetic resources and the concept of harm. *Journal of Law, Medicine & Ethics*, 35(3), 396-411; Tobin, above n 61, 293 *et seq.*

111 Tobin, above n 61, 294.

PART III
Social, Economic and Political Aspects of Food and Intellectual Property

Chapter 6
Why Didn't an Equivalent to the US Plant Patent Act of 1930 Emerge in Britain? Historicizing the Boundaries of Un-Patentable Innovation

Berris Charnley

Introduction

In 1906 the Royal Horticultural Society hosted the Third International Conference on Hybridisation and Cross Breeding. The meeting is remembered now as a triumph of the Mendelian school, whose leader, William Bateson, used the occasion to put into public circulation his new name for the discipline: genetics. Less widely remembered is that the meeting also hosted a session bringing together plant breeders to discuss another, related, matter: whether they could use copyright law to protect their IP in the novel varieties that, they reckoned, would surely arise ever more abundantly thanks to the work of the Mendelians. So plant IP was on the minds of plant breeders in Britain from the early years of the twentieth century. And yet over the succeeding decades there would never emerge anything comparable to the US Plant Patent Act of 1930. Why not? This chapter will offer a preliminary answer, emphasizing the extent to which different paths of agricultural development, and experiences of the Great War, shaped the social – political and moral – contexts of plant breeding in each country. Furthermore, in conclusion, the chapter will consider the relevance of this history for contemporary discussions over the boundaries of patentable subject matter.

The US Plant Patent Act of 1930 (PPA) is a well-worn passage point in the history of patenting life.[1] Broadly speaking, there are three sets of reasons on the

1 On the foundation of the PPA, the key historical articles are probably Kevles, D.J. 2007. Patents, Protections, and Privileges: The Establishment of Intellectual Property in Animals and Plants. *Isis*, 98, 323-331; Kevles, D.J. 2008. Protections, privileges, and patents: Intellectual property in American horticulture, the late nineteenth century to 1930. *Proceedings of the American Philosophical Society*, 152(2), 207-217; Kevles, D.J. and Bugos, G.E. 1992. Plants as intellectual property: American practice, law, and policy in world context. *Osiris*, 7, 75-104; Fowler, C. 2000. The Plant Patent Act of 1930: A sociological history of its creation. *Journal of the Patent and Trademark Office Society*,

table as to why the PPA came into place when and as it did in the United States. We might call these three sets of causes of the act's existence geographic, scientific and social.[2] Into the first of these categories we can place contemporary arguments for plant patents which hinged on the rapid geographical expansion of American plant breeding: from east to west and back east again.[3] Across these widely dispersed and rapidly expanding markets, informal community-based chastisement would, it was feared, hold little sway. Accordingly, breeders like Luther Burbank, on the west coast, argued that they needed protection from variety thieves who could smuggle plants over to east coast markets, and vice versa for the breeders on the east coast.[4] A second set of contemporary arguments in favour of the PPA focused on advances in scientific theory, and in particular Mendelism and pure line theory. These new theoretical developments were, some thought, capable of clearing up the problem of how to argue that breeders constructed new varieties, in a similar manner to chemists and engineers who, in the period, were becoming increasingly adept at manipulating molecules and machines. Cary Fowler, an activist turned historian of patenting life, voices most strongly a concern that runs throughout the historiography: did Mendelism actually change plant breeding in this early period? Fowler is at pains to dismiss the view that it did, pointing out that Mendelism had little to do with plant breeding before the 1930s.[5] On Fowler's interpretation of the influence of science, Mendelism and pure line theory had a more subtle didactic and sociological role to play in arguments put forward by the pro-plant patent

82(9), 621-644. See also Moser, P. and Rhode, P.W. 2011. *Did Plant Patents Create the American Rose?*, NBER Working Paper No. 16983. Cambridge, MA: National Bureau of Economic Research for a revisionist account of the PPA's economic impact on rose breeding.

2 Of course, the key sources given above on the foundation of the PPA rely on a sophisticated analysis of the combination and interplay between these factors this list is intended to be used instrumentally rather than as a comprehensive diagnosis of the American case or its historiography.

3 On the westward expansion of American agriculture see Olmstead, A.L. and Rhode, P.W. 2008 *Creating Abundance: Biological Innovation and American Agricultural Development*. New York: Cambridge University Press, chapters 2 and 3, and on westward expansion and the circulation of plant varieties back to the east coast see Pauly, P.J. 2007. *Fruits and Plains: The Horticultural Transformation of America*. Cambridge, MA: Harvard University Press, especially chapters 5, 6 and 7.

4 See Kevles, above n 1 ('Patents, Protections, and Privileges'), 326 and Fowler, above n 1, 623.

5 Fowler, above n 1, 624-626. The idea that early Mendelism had little to offer plant breeders draws heavily from Mayr, E. 1982. *The Growth of Biological Thought: Diversity, Evolution and Inheritance*. Cambridge, MA: Harvard University Press. See also Kevles, above n 1 ('Protections, Privileges, and Patents'), 208-209 and ('Patents, Protections, and Privileges'), 325.

lobby.[6] These theories acted as a scientific fig leaf of respectability around what had previously been thought of as the *art* of plant breeding. Finally, the people who made these arguments, and the way in which they made them, were important ingredients in the mix of causes that led to the PPA. Paul Stark, of Stark Brothers' Nurseries, was loud and vociferous in his well-connected campaign to have the PPA enacted, one to which the figures of Luther Burbank and Thomas Edison lent much weight (and in Edison's case ample opportunity for comparison of breeders to engineers).[7] In other words there were strong social factors – relating to coordination and access – behind the instantiation of the PPA.

One way of developing this lively interpretive debate is to ask, if the PPA was enacted in America, why there but not elsewhere? Posing this question, in the first place, throws light on what was happening in Britain. Secondly, it helps in formulating new questions about the American legislation that was enacted and contemporary legislation being considered today. The absence of IP rights in British plants, prior to the introduction of the *International Union for the Protection of New Varieties of Plants* (UPOV), has received little attention.[8] However, this picture is changing; set within a general trend towards focusing on historical patterns of patenting, but in conscious contrast to it, recent work in the history of economics on patent-free innovation has suggested that plant variety innovation occurred rapidly across Europe even though plant patents were

---

6 On purity, pure lines and the production of industrial products see Bonneuil, C. 2008. Producing identity, industrializing purity: Elements for a cultural history of genetics, in *A Cultural History of Heredity, IV*, edited by S. Müller-Wille and H.-J. Rheinberger. Berlin: Max Planck Institute for the History of Science Preprint Series. See Sherman, B. and Pottage, A. 2007. Organisms and manufactures: On the history of plant inventions. *University of Melbourne Law Review*, 31(2), 539-568 on the types of disclosure and specification necessary for obtaining a plant patent and on the importance of specification to modern patent systems. More generally see Mario, B. 2006. Patent republic: Representing inventions, constructing rights and authors. *Social Research*, 73(4), 1129-1172.

7 See Fowler, above n 1, 634; Moser and Rhodes, above n 1, 3.

8 There is a small amount of work on this subject, a passage or two from historian of science Palladino, P. 1990. The political economy of applied science: Plant breeding research in Great Britain, 1910-1940. *Minerva*, 28(4), 449-450, and Palladino, P. 1996. Science, technology, and the economy: Plant breeding in Great Britain 1920-1970. *Economic History Review*, 49(1), 130-131, another small section in the more strictly legal history from Llewellyn, M. and Adcock, M. 2006. *European Plant Intellectual Property*. Oxford: Hart, 141-143, an unpublished PhD thesis, Rangnekar, D. 2000. 'Innovative Appropriation: The Role of Law, Economics and Science in Transforming Plants into Products – A Case Study of Wheat Breeding in the UK'. Unpublished PhD Thesis, Kingston University, and a short paper from agricultural economist, Berlan, J.-P. 2001. The political economy of agricultural genetics, in *Thinking about Evolution: Historical, Philosophical, and Political Perspectives*, edited by R.S. Singh et al. Cambridge: Cambridge University Press.

unavailable.[9] We also know more about the ways in which British plant breeders sought to protect their new varieties when IP rights were unavailable to them.[10] This chapter supplements these analyses and develops a new line of enquiry into the history of patenting life by offering an explanation for the absence of plant patents in 1930s Britain. The chapter also has a secondary and more self-conscious aim; to link this type of historical reconstruction with the contemporary debates in which the current humanities turn to innovation and IP studies has been motivated.

Viewed from this angle, Britain is an ideal case study of a country in which plant patents were not enacted. We begin with the views of a group of plant breeders at the Royal Horticultural Society's Third International Conference on Hybridisation and Cross Breeding. At their session there was mixed enthusiasm for IP rights and even among those who were enthusiastic there was little coherence. The diary of an American breeder touring Britain two years later, in 1908, then gives us a window on to Edwardian British plant breeding. A new part of this scene was publicly funded plant breeding, which, as an aspect of national agricultural revival, was becoming a popular cause. In the next part of the chapter we turn, via the British agricultural experience of the Great War, to post-war public funding of plant breeding. One indicative highpoint of these plans was the mid-1920s release of Yeoman II; the country's first fully publicly funded wheat variety, distributed by a new National Institute of Agricultural Botany. By the mid-1920s, just as government seed programs were ending and calls for the PPA were gaining ground in America, publically funded plant breeding monopolized the British context, and the plant breeders at commercial firms were quite openly snubbed. In closing, the chapter turns to its secondary aim asking what value might be drawn from reconstructing the particular notions of morality, equity and nation invoked by early twentieth-century plant breeders.

9 For more on areas of innovation without patents, see Nuvolari, A. 2004. Collective invention during the British Industrial Revolution: The case of the Cornish pumping engine. *Cambridge Journal of Economics*, 28(3), 347-363. On rates of patent-free plant innovation see Boldrin, M. and Levine, D. 2008. *Against Intellectual Monopoly*. Cambridge: Cambridge University Press, 57-61.

10 On plant breeders' alternative strategies to patenting, see Charnley, B. 2012. 'Agricultural Science, Plant Breeding and the Emergence of a Mendelian System in Britain 1880-1930'. Unpublished PhD thesis: University of Leeds, 2012, chapter 3 and Charnley, B. and Radick, G. 2010. Plant breeding and intellectual property before and after the rise of Mendelism: The case of Britain, in *Living Properties: Making Knowledge and Controlling Ownership in Modern Biology*, edited by D. J. Kevles et al. Berlin: Max Planck Institute for the History of Science Preprint Series.

Plant Breeding in Britain before the Great War

'"Copyright" for the Raisers of Novelties'

The Royal Horticultural Society (RHS) was founded in the 1830s and initially represented the interests of aristocratic gardeners. However, after near bankruptcy in the 1880s, the society was set on a new course by Sir Daniel Morris, a man who had made his reputation administering colonial agriculture in the West Indies. To rescue the society he fostered links with commercial plant breeders and academic botanists.[11] By the close of the century the society was a key feature of Edwardian horticultural life. In 1897, William Bateson, in his role as secretary for the Royal Society's evolution committee, approached the RHS's members for records of their breeding activities. Bateson wanted to use these records to analyse heredity.[12] His efforts had mixed results, a stud book of orchids was produced, with the help of Bateson's long-term collaborator and RHS insider, Charles Hurst, but many breeders remained silent.[13] One wholly positive upshot, however, of Bateson's developing relationship with the society was a series of international conferences on plant breeding techniques which started in 1899.[14] The second conference, hosted by the New York Horticultural Society in 1901, was a key moment for Mendelian aspirations, but it is the third conference, held in London and Cambridge in 1906, that is the most well-known of the three events.[15] The proceedings were recorded

11 See Morris's speech on the history of the society in Wilks, W. 1907. *Report of the Third International Conference on Genetics*. London: Spottiswoode, 56. Morris became an enthusiastic Mendelian, using the theory in his sugar cane work which he was demonstrating at the conference that year.

12 For a recent biography of Bateson see Cock, A.G. and Forsdyke, D.R. 2009. *Treasure Your Exceptions: The Science and Life of William Bateson.* New York: Springer. On Bateson's early career see Olby, R. 1989. The dimensions of scientific controversy: The biometric-Mendelian debate. *British Journal for the History of Science*, 22, 299-320 and Kevles, D.J. 1980. Genetics in the United States and Great Britain, 1890-1930: A review with speculations. *Isis*, 71(3), 441-455.

13 See Hurst, R. 1949. The R.H.S. and the birth of genetics. *Journal of the Royal Horticultural Society*, 74, 380-381.

14 On Bateson's relationship with the RHS see Punnett, R. 1950. Early days of genetics. *Heredity*, 4, 1-10, Hurst, above n 13, and two short articles, both timed to celebrate the Mendelian 'rediscovery', from Olby, R. 2000. Horticulture: The font for the baptism of genetics. *Nature Reviews Genetics*, 1, 65-69 and Olby, R. 2000. Mendelism: from hybrids and trade to as science. *Comptes Rendus de L'Academie des Sciences. Serie III, Sciences de la vie*, 323(12), 1043-1051.

15 See Radick, G. 2013. The professor and the pea: William Bateson and the case for Mendelian genetics as applied science, 1902-1922. *Studies in the History and Philosophy of Science Part A*, in press 2013 for a detailed exegesis of Bateson's paper and his audience's early responses to Mendelism in New York.

and published by Reverend W Wilks, the society's secretary, in the following year, adopting Bateson's new word, 'genetics', for the report's title.

At the fifth session of the third conference, '"Copyright" for the Raisers of Novelties', British commercial plant breeders (and two foreign academic breeders) came together to discuss the issue of plant breeding IP. Opening the session, Mr George Paul VMH, a commercial rose breeder, held the floor.[16] He wanted to coax the society into officially passing a resolution on the desirability of copyright for plant breeders. Paul felt some sort of right was necessary as recompense, 'for all the risk and labour, added to the observations and experience which have taken the best years of one's life to amass'.[17] Paul noted the absence of many breeders at the conference and suggested this was because, 'these Gentlemen do not like to tell us or to show, what they have done in their experiments, because when once their knowledge becomes public, they have not the slightest chance of receiving any pecuniary reward for their labours'.[18] As a result of this move to secrecy, Paul complained, their 'invaluable' knowledge was inaccessible to other breeders, as it had been to Bateson. On Paul's view, some version of the patent bargain, whereby innovators were rewarded with state protection in exchange for sharing their new innovation with society, would benefit the plant breeding community. Professor Hansen, probably Emil Chr. Hansen, Head of the Department of Physiology at the Danish Carlsberg Laboratory, 1879-1909, summarized the group wisdom that, 'in law, a seedling is regarded as the gift of God, and it would be hard to patent that'. However, Hansen shared Paul's desire for 'some law fashioned ... [to] give a bonus to the man who does such skilled and valuable work'.[19]

Where precision was a problem, this was not considered, by at least one member of the discussion, to be a problem of exact manipulation, or even constancy through generations but rather one of identification in the first place. In the discussion that followed Paul's remarks, the unnamed chairman of the session noted, 'surely our discussions to-day show what a very great difficulty there would be in enforcing such law, because we have gentlemen from all parts of the world maintaining that a thing is new, and others, equally capable, maintaining that it is old'.[20] As a result of the apparent proliferation of varieties that had occurred in the previous century it was often difficult to tell what was really new. Sometimes this was the result of an innocent taxonomical mistake, as, 'many raisers may be occupied in cultivating the same class of a plant, and two or three of them might get something very similar, at the same moment'.[21] Often though, such confusion

16 VMH, the Victoria Medal of Honour, was awarded by the RHS Council as a special honour by the Society.

17 Wilks, above n 11, 474.

18 Ibid.

19 Ibid.

20 Ibid., 475.

21 Ibid.

was the result of direct piracy of either a varieties' name, or the variety itself: this was a problem that scientific precision was, apparently, unable to broach.[22]

There was considerable sympathy for Paul's suggestion from one German attendee to the session, Professor Wittmack. The professor described how in Germany, 'we are protected by the laws of our agricultural society, which is a very great society. A man who has bought a specimen of a novelty from the raiser, dare not himself sell it.' Wittmack continued by wondering that, 'surely a society like the RHS could do the same, and they should expel a man who did not do as they wished'. However, some of Paul's fellow British breeders were rather less sure about the possibility or even the benefits of protection, especially to the trade as a whole. Mr James Douglas suggested, 'It appears to me that protection would cut both ways … Mr Paul propagates other people's roses and other people propagate his. It is perfectly fair … We do not want to restrict the sale of plants but increase it.' Finally, the chairman concluded that 'the point raised by Mr Paul is a most interesting one, but there are evidently two sides to it, as to most other things, and, unless there were a very decided majority in favour of it, I do not think it would be wise for us to move in the matter'.[23]

The problem in Britain, for those who wanted legislation, was that there was no consistent majority line around which breeders conformed, instead there were a plurality of views on subject. This disarray might be compared to the US situation, where Paul Stark lobbied very effectively for the PPA, with wide support from the American Seed Trade Association and the American Breeders' Association and, of course, his congressman.[24] As Margaret Llewellyn and Mike Adcock have noted about the genesis of UPOV rights, 'the development of the right would not have occurred had it not been for the widespread commitment of the breeding community'.[25] If at this early point in the century there was little consensus amongst plant breeders, what sort of shape was British plant breeding in? To cast more light on this question we can turn to the diary of George Shull, recorded as he toured British plant breeders in 1908.

---

22 Charnley and Radick, above n 10, discusses one case of variety appropriation in the nineteenth century and the continuing problems this caused for Mendelian breeders concerned with protecting the reputations of their varieties. On the total number of new varieties in the nineteenth and twentieth centuries, there are several collections of seed trade catalogues at the Museum of English Rural Life and the National Institute of Agricultural Botany's archives. A quantitative analysis of these records might shed useful light on rates and weight of innovation amongst British plant breeders.

23 Wilks, above n 11, 475.

24 Indeed, support from the Agricultural Seed Traders Association continued even after the restriction to asexual plants, and not those reproduced by seed, was self-imposed on the campaign by Stark, see Fowler, above n 1, 635.

25 Llewellyn and Adcock, above n 8, 143.

George Shull's Tour of British Plant Breeders

While visiting the 1908 Dublin meeting of the British Association for the Advancement of Science, George Shull included in his trip a tour of British plant breeders.[26] His choices as to which plant breeders to visit and the work he found being carried out are both instructive. The picture we get of the British plant breeding community from Shull's diary is of one populated largely by commercial breeders, many of them patriarchal family firms, few of whom seem to have been undertaking high levels of breeding, much less Mendelian breeding.[27]

Shull's trip began on 8 August 1908. At the end of the month he visited the Thomas Rivers & Sons Nursery at Sawbridgeworth, near Cambridge. At the nursery he found a great deal of impressive production, achieved through the use of dwarfed fruit trees, 'Many tons of apples and other fine fruits are sent to market every year and great quantities of nursery stock are sold.' However, Shull found little evidence of breeding, and that, '[t]he breeding work has been necessarily simply a side issue and has been carried on largely for the sake of advertisement'.[28] At Thomas Laxton's nurseries the picture was pretty much the same, the sons of the great man had 'continued the work, always as an adjunct to an ordinary nursery business'.[29] The exception to this picture amongst the commercial breeders Shull visited was the Suttons' nurseries. Here, the sons of Arthur Sutton, who like the old man, were close friends of William Bateson, were understandably keen on the new Mendelian breeding.[30]

When Shull finally got to visit an overworked Bateson, arguably the high point of his trip, it was at Grantchester, near Cambridge, at Bateson's family home, where he was conducting all of his research. Mendelism, despite the high praise of the RHS, still did not have an institutional home in 1908; it was literally a

26 For more on Shull, who was an important part of the hybrid maize story and the American reception of pure line theory, see Bonneuil, above n 6. Shull was very impressed by the Reverend Wilks's report of the RHS's conference and state of the art Mendelism in Britain, the report was reviewed by Shull for the *Botanical Gazette*, see Shull, G. 1908. Review: Conference on genetics. *Botanical Gazette*, 46, 60-61.

27 For more on the firms that did breed see Olby, above n 14 ('Horticulture' and 'Mendelism').

28 Shull, G. 'Notes on European trip of Geo. H. Shull, Aug. 12, 1908 to Dec. 1908', George Harrison Shull Papers. American Philosophical Society, Mss. B Sh92, 27. Many thanks to Marsha Richmond for lending me her transcript of this document.

29 Ibid, 44. Thomas Laxton, like Thomas Rivers and Arthur Sutton were great fans of, and sometime correspondents with, Charles Darwin, who courted them much as Bateson courted the RHS, for their experimental results.

30 Bateson shared his new varieties with the family in return for raw materials and breeding data, see the long-running correspondence between the Suttons and Bateson in the Bateson Letters Collection at the John Innes Centre Archive.

cottage industry.[31] During his trip Shull did not visit any public plant breeding stations; there would not be one in existence in Britain until 1912.[32] In contrast, the commercial nurseries that he visited were flourishing, and we could add to this list Carters, Veitch and Gartons, all of which had recently moved into international markets including America and Australia.[33] However, the amount of breeding that was actually conducted, at least on Shull's sampling, is less than one would have expected. The great British seed houses and nurseries of the nineteenth century, with a few notable exceptions, were increasingly involved in seed dealing rather than plant breeding.[34]

The Great War and the New Mendelian System

In the same year as Shull's tour, the fame of another sort of plant breeder was beginning to grow. This new class of academic plant breeders had their own hero: Rowland Biffen.[35] While Shull was making his trip around Britain, Biffen was made chair of Agricultural Botany at the Cambridge University's Agricultural

31 On Mendelism as a cottage industry and on the early years of Mendelism at Cambridge, see Richmond, M.L. 2001. Women in the early history of genetics: William Bateson and the Newham Collage Mendelians, 1900-1910. *Isis*, 92(1), 55-90 and Richmond, M.L. 2006. The 'domestication' of heredity: The familial organization of geneticists at Cambridge University, 1895-1910. *Journal of the History of Biology*, 39(3), 565-605 and Olby, above n 12.

32 Compare Shull's tour of British plant breeding with later trips by wheat breeder to the New Zealand Government, Otto Frankel, which focused almost exclusively on public stations. See Frankel, O. 1936. Frankel's tour, in *Reports of the Wheat Research Institute 1935-36, volume IV*, New Zealand Archives, Wellington. Series: 20502, 139-186.

33 See Carters's advert in the front of the *Journal of the Board of Agriculture* from 1906 to 1910.

34 This mixture of breeding and dealing echoes Fowler's view of the American case: see Fowler, above n 1.

35 See Engledow, F.L. 1950. Rowland Harry Biffen 1874-1949. *Obituary Notices of Fellows of the Royal Society*, 7, 9-25 for a fairly revealing obituary. Notices also appeared in *Nature*: Engledow, F.L. 1949. Obituaries: Sir Rowland Harry Biffen F.R.S. *Nature*, 164, 305 and *The Times*: 'Obituary: Sir Rowland Biffen: Pioneer of Agricultural Botany', *The Times*, 15 July 1949: 9d. See also Palladino, P. 2002. *Plants, Patients and the Historian: (Re)Membering in the Age of Genetic Engineering*. Manchester: Manchester University Press, 81-92 for a biography. For more salacious background detail, including Biffen's marriage to the younger sister of his first love; and his in-laws and wife Mary's prize winning sweet pea breeding work, see Taylor, J.M. and Wilkinson, S. 2008(?). Miss Evelyn Hemus. *Horthistoria.* [Online]. Available at: http://horthistoria.com/?p=217 [accessed: 25 October 2012] and Taylor, J.M., Wilkinson, S. and Hammett, K. 2009(?). Miss Hilda Hemus. *Horthistoria.* [Online]. Available at: http://horthistoria.com/?p=219 [accessed: 25 October 2012].

Department. Although his chair was initially funded by the Worshipful Company of Drapers, Biffen's research was increasingly funded by the Board of Agriculture and after 1910, another government body, created by then chancellor of the exchequer, David Lloyd George, with the aid of Winston Churchill, to aid rural reconstruction: the Development Commission.[36] The Commission was initially charged with managing a £1million fund, although in subsequent years this money grew substantially. A.D. Hall, who was at one time a collaborator in research with Biffen, was appointed head of the Commission and his influence shaped its activities over the next seven years.[37] Biffen was personal friends with many of the commission's members and served, along with Bateson, on its sub-committees on several occasions.[38]

In 1910 Bateson came to be in control of his own research institute, the John Innes Horticultural Research Institute, although the institute did not receive significant public funding.[39] At the same time, using the public funds that he increasingly commanded, as a Whitehall insider, Biffen established a Plant Breeding Institute at Cambridge University. Several students and many plant

---

36 For more on the Development Commission, its foundation and aims see Olby, R. 1990. Rôle de l'agriculture et de l'horticulture britanniques dans le fondement de la génétique expérimentale, in *Histoire de la Génétique: pratiques, techniques et theories*, edited by J.-L. Fischer and W.H. Schneider. Paris: ARPEM and Olby, R. 1991. Social imperialism and state support for agricultural research in Edwardian Britain. *Annals of Science*, 48(6), 509-526. For more on the Drapers' Society see Gooday, G. 1998. The premisses of premisses: Spatial issues in the historical construction of laboratory credibility, in *Making Space for Science: Territorial Themes in the Shaping of Knowledge*, edited by C. Smith and J. Agar. Basingstoke: Macmillan. Winston Churchill enjoyed at least three political careers; he was Prime Minister twice and switched parties three times. In 1909, along with David Lloyd George, he was a young liberal MP, very much involved in social reform.

37 Biffen and Hall collaborated from roughly 1900 to 1906 through a group called the Home Grown Wheat Committee, sponsored in turn by the National Association of British and Irish Millers see Biffen, R. and Humphries, A.E. 1907. The improvement of English wheats. *Journal of Agricultural Science*, 2, 1-16 and Biffen, R. and Humphries, A.E. 1911. Home-grown wheat committee. *The Times*, 28 August, 4a. See also Dale, H.E. 1956. *Daniel Hall: Pioneer in Scientific Agriculture*. London: John Murray for a biography of Hall.

38 For Biffen's work on the Advisory Committee on Agricultural Science to the Development Commission see Development Commission. 1918. [Cd. 118] *Eighth Report of the proceedings of the Development Commissioners for the year ended 31st March 1918*. London: His Majesty's Stationary Office, 2.

39 The John Innes Horticultural Research Institute of which Bateson became director owed more in the form of its foundation to the benefactor grants of the previous century which had established the Rothamsted station and Royal Agricultural College. See Olby, R. 1989. Scientists and bureaucrats in the establishment of the John Innes Horticultural Institution under William Bateson. *Annals of Science*, 46(5), 497-510; Olby, above n 36 ('Rôle de l'agriculture et de l'horticulture' and 'Social Imperialism and State Support').

varieties circulated between Biffen and Bateson's institutes.[40] However, the institutional system that Biffen and his close friend Bateson, along with several other Cambridge- and Whitehall-based agricultural improvers were constructing, ostensibly remained inactive for the duration of the war.[41] From 1916, Biffen worked at the Food Production Department: a body formed for the emergency nationalization of British agriculture. While at the unit Biffen further strengthened his connections to government at the same time as government was strengthening its control over British agriculture.

Britain was largely unprepared for the war and apparently came remarkably close to mass starvation. On the one hand British agriculture, after years of depression, was ill equipped to feed the population, on the other hand German U-boats were cutting off the majority of provisions from the merchant navy. Nationalization, forced ploughing and government-backed fixed wheat prices where all part of a package of measures introduced to increase average calorific intake. The figurehead of these measures, who by his own reckoning played no small part in winning the war, or least not losing it, was Thomas Middleton, former colleague at Cambridge, and now superior of the seconded Biffen at the Food Production Department.[42] By the end of the war, with David Lloyd George as Prime Minister, Biffen, Bateson and many other Mendelians were every bit as connected to the legislature, through the Food Production Department, Board of Agriculture and Development Commission, as their American counterparts.

The Plans for Post-War Reconstruction

At the close of the Great War, A.D. Hall, who had left the Development Commission and was now near the top of the Board of Agriculture's civil service apparatus, wrote to Bateson asking his advice on post-war reconstruction.[43] Bateson's reply

40 See, for example, 'Rowland Biffen to William Bateson, 27 November 1912', William Bateson Letters, John Innes Centre Archives, Mss 2224.

41 For more on the systematic ambitions of this group see Charnley, B. 2013. Managing knowledge in 'systematised plant breeding': Mendelism and British agricultural science, 1900-1930, in *Knowledge Management and Intellectual Property: Concepts, Actors and Practices from the Past to the Present*, edited by G. Dutfield and S. Arapostathis. Cheltenham: Edward Elgar, in press.

42 See Middleton, T. 1923. *Food Production during War*. Oxford: Clarendon Press and for a history of technology oriented revision of this account see Dewey, P.E. 1980. Food production and policy in the United Kingdom, 1914-1918: The Alexander prize essay. *Transactions of the Royal Historical Society*, 30, 71-89 and Dewey, P.E. 1989. *British Agriculture in the First World War*. London: Routledge, in which Dewey emphasizes women in the agricultural labor force, new technologies and a range of other factors than Middleton's interventions. See also Biffen's obituary of Middleton, Biffen, R. 1943. Obituaries: Sir Thomas Middleton K.C.I.E., K.B.E., C.B., F.R.S. *Nature*, 152, 97-98.

43 For more on Hall's plans for post-war reconstruction see Hall, A.D. 1916. *Agriculture after the War*. London: John Murray.

was complex, but it reveals a great deal about Mendelian plans. Bateson was wary of any government funding for his own institute, feeling that the independence of agricultural research from practical aims should be safeguarded.[44] However, he did see a role for government support. First, he suggested that Hall should organize for a 'liaison officer' to be employed by the Board of Agriculture as a go-between for Mendelians and the rest of the plant breeding community. Bateson, writing in 1917, felt the Mendelians, if they were to aid post-war agricultural reconstruction, needed someone to explain their work and encourage other breeders to make use of it themselves.[45] Secondly, he felt that Mendelians needed a way to distribute their new varieties:

> 2. It is desirable that we should be in close association with an institution able to grow our things on a farm scale. Hitherto our efforts to get this done have been failures… It is unlikely that, except under strong pecuniary inducement, anyone will take the special care which such work demands, when he feels that the chief credit for success will go to someone else. In work of this kind we are in the hands of the grower, and a bad result can always be ascribed to an inherent peculiarity of the material. Perhaps the Cambridge Svalof (*sic*) Station will cover this want.[46]

---

44 Bateson's view on the danger of public funding was based on his perception of American land grant university research stations. He believed near sightedness had made the work they conducted 'trivial and unfruitful', see *William Bateson to A. D. Hall, 23rd February 1918*, William Bateson Letters, John Innes Centre Archives, Mss 2148, 2. See also Bateson's criticisms of American research stations at the Portsmouth meeting of the agricultural subsection of the British Association for the Advancement of Science, Bateson, W. 1912. Chairman's Address to the Agricultural Sub-Section. *Report of the Eightieth Meeting of the British Association for the Advancement of Science: Held at Portsmouth in August and September 1911*, Sub-Section M–Agriculture, 588-589. For a detailed analysis of Bateson's views on the complex alignment of pure and practical research that should be held as ideal, and the divergent historiographical traditions of representing those views see Radick, above n 15.

45 Bateson's suggestion that Mendelism was complicated enough to need explaining fits with Fitzgerald's views on the increasing complexity of Mendelian plant breeding Fitzgerald, D. 1991. Beyond tractors: The history of technology in American agriculture. *Technology and Culture*, 32(1), 114-126. However, it is interesting that this is not the only view Bateson held on the subject. Like Biffen, he also often extolled the simplicity of the theory and its self-evident application to plant breeding, see for example Bateson, W. 1902. *Mendel's Principles of Heredity: A Defence.* London: Cambridge University Press, 208. It seems what Bateson was more interested in was a publicity officer. His choice of candidate supports this second view. He put forward W.O. Backhouse, an academic breeder from a longstanding and respected family of commercial daffodil breeders.

46 *William Bateson to A. D. Hall, 23rd February 1918*, above n 44, 3.

Bateson was obviously sceptical about placing the job of multiplying new Mendelian varieties in the hands of the commercial seed trade. He felt that this was a job for government, especially a government which was now emboldened in its desire to intervene in agriculture as a result of its experiences during the war. The Cambridge Svalöf station which Bateson referred to was Rowland Biffen's Plant Breeding Institute, which Biffen was hoping to expand through a new plan for government-backed and partly nationalized seed distribution.

Biffen shared Bateson's view that the proper role of government should be as an instrument to promote new Mendelian varieties, although along with Bateson he was by no means prone to socialist thinking.[47] After the war, Biffen began working with Lawrence Weaver, his superior in the wartime Food Production Department, which was led by Middleton, on establishing a National Institute of Agricultural Botany (NIAB). Biffen's plan for the new institute drew much from the arrangements at Sweden's Svalöf station.[48] He envisaged NIAB as being an ancillary unit to his own Plant Breeding Institute, charged with commercializing new Mendelian varieties. He expected NIAB to be financially self-sufficient from the sale of new varieties. In the first ten years of the institute's operation (1919-29) Biffen's plans were largely, if not entirely, realized. Even in 1929, when Biffen's plans for self-sufficiency failed, and the institute sought to scale back its operations, more government funding was allocated to stabilize the change.[49]

---

47 Although there was undoubtedly a socialist element in these circles; Sydney Webb, the renowned Fabian and socialist thinker, was a member of the Development Commission for the first three years in which it ran, see Development Commission. 1911. [Cd. 199] *First Report of the Proceedings of the Development Commissioners for the Periods from 12th May 1910 to 31st March 1911.* London: His Majesty's Stationary Office for a list of committee members and the first in a series of yearly reports. See also Hale, P. 2010. Of mice and men: Evolution and the socialist Utopia. William Morris, H. G. Wells, and George Bernard Shaw. *Journal of the History of Biology*, 43(1), 17-66 on the diverse nature of British socialists during the revival of the 1880s and afterwards.

48 See the eight-page plan for the new institute from Biffen, *Rowland Biffen to Lawrence Weaver, 3 December 1917*, Correspondence 1916-1918. Archives of the National Institute of Agricultural Botany, the copious material on Svalöf and the several trips made there by NIAB staff in the institute's boardroom library.

49 This realignment was largely orchestrated by Edwy Sloper Beaven, see *Memorandum on Future 'Crop Improvement Policy of the NIAB.* Council Paper no. 65. Archives of the National Institute of Agricultural Botany and Palladino, P. 1994. Wizards and Devotees: On the Mendelian Theory of Inheritance and the Professionalization of Agricultural Science in Great Britain and the United States, 1880-1930. *History of Science*, 32, 409-444 for biographical details. On Beaven's view, realized from around 1928, the institute should maintain its monopoly on the innovation of new varieties but let the seed trade take up the task of multiplication. Despite Bateson's hopes, the institute drew back from its own commercial production.

The context in which NIAB was established is encapsulated in Britain's post-war reconstruction committees, known as the Selborne and Linlithgow reports.[50] Biffen appeared before the Selborne Committee, which reported in 1918, and on which A.D. Hall was a commissioner, to give evidence as an expert on the state of the wheat growing industry in Britain and around the world. In this international context, the committee's conclusions made obvious the need, they felt, for personal sacrifice to maintain national security. At the heart of their recommendations was the notion that,

> The State must, in short, take every means in its power to give confidence and a sense of stability to landowner farmers and agricultural labourers. It must then tell those classes exactly what is expected of them, and appeal to their highest instincts of patriotism to put personal predilections aside and unite to carry out a policy, on the success of which the safety of their nation may someday depend.[51]

The grandeur of these national plans, which included powers to requisition and repurpose land used frivolously for 'game or games', meant that attempts to secure personal rights, even to one's literal property, could be tantamount to treason. Although many of the Selborne committee's recommendations were never enacted, the creation of a National Institute governing the supply of seed was exactly of a piece with these plans.

Yeoman II: the First Publicly Funded Wheat Variety

In 1924, the same year that free seed programs were ending in the US, Biffen was preparing the release of a new Mendelian variety, Yeoman II.[52] This high-profile

---

50 The Selborne Commission was instigated by the Board of Agriculture and Fisheries under the terms, 'To consider and report on the methods of effecting an increase in the home production of food supplies, having regard to the need of such an increase in the interests of national security'. See Board of Agriculture and Fisheries. 1918. [Cd. 8506]*1917-18: Report of the Agricultural Policy Sub-Committee of the Committee on Reconstruction, appointed to consider the methods of effecting an increase in the home-grown food supplies, Part I.* London: His Majesty's Stationary Office, 1.

51 Board of Agriculture and Fisheries, above n 50, 16. Biffen and NIAB received more direct praise and enthusiasm, with less grandeur in 1923 when the Linlithgow Commission, appointed as a successor to the Selborne Commission, under the renamed Board of Agriculture, reported on the wheat industry. See Ministry of Agriculture and Fisheries. 1923. [Cmd. 1971] *Departmental Committee on Distribution and Prices of Agricultural Produce: Interim Report on Cereals, Flour and Bread, [13th Dec. 1922-9th Oct. 1923].* London: His Majesty's Stationary Office, 86-98.

52 On the end of American free seed programs see Fowler, above n 1, 632. On the release of Yeoman II see Biffen, R. 1924. The new wheat Yeoman II. *Journal of the Ministry of Agriculture*, 31, 509-12 and Parker, W.H. 1923. Report of trials of Yeoman II wheat 1922-23. *Journal of the National Institute of Agricultural Botany*, 1, 2-8.

moment for publicly funded schemes stands in marked contrast to the increasing dominance of commercial firms in American plant breeding.[53] In autumn that year Yeoman II was officially made available for sale. The variety was produced by Biffen at the PBI and multiplied and distributed by NIAB. Announcements were made in the *Journal of the Board of Agriculture*, the agricultural press and even in the *Times* and popular press.[54] Seeds were available, at a fixed price, in sealed sacks, directly from NIAB, who were keen to point out in a catalogue they produced for the release, 'none but this will be genuine'.[55] In the following year several reports of the success of the release (rather than the variety) also appeared.[56] As to the success of the new institutes at which Biffen worked, the Prince of Wales voiced this approval in his 1926 presidential address to the British Association for the Advancement of Science: 'At the plant-breeding institute at Cambridge, Sir Rowland Biffen has provided several new wheats, of which two are generally grown throughout the country; the extra yield and value of these wheats must already have more than repaid the whole expenditure on agricultural research since the institute was founded.'[57] The Prince obviously felt that the expense of public breeding was more than justified by the benefits brought to the nation. The fact that NIAB was actually struggling to realize self-sufficiency at this point should not detract from the importance of perceptions of its success promoted by, amongst others, the Prince of Wales and the Prime Minister.[58] In their time and place Biffen and his work were quite as famous and influential as Paul Stark, if not Luther Burbank. But unlike either of these figures, Biffen was almost entirely reliant on public funds for research income.

---

53 See Lewontin, Levins and Burlan's work on this subject, in a series of *Monthly Review* articles and elsewhere which can be located from the bibliography of Lewontin's work given in Singh et al., above n 8, 583-589.

54 1924. New English wheat Yeoman II available for sowing. *The Times*, 28 July, 18b.

55 *Yeoman II*, Papers of A. D. Hall, John Innes Centre Archives. Mss 188-8-66.

56 See for instance, 1925. Yeoman II seed wheat: 2480 Quarters Marketed. *The Times*, 6 January, 8e.

57 The contents of this presidential address were (as in most years) republished in *Science*, no doubt increasing Biffen's fame internationally. See, 1926. The Presidential address. *Science*, 64, 146. The Prince of Wales refers to two wheat varieties, Little Joss and Yeoman, released in roughly 1910 and 1916 respectively by Biffen through an ad-hoc system of distribution. While both of these varieties were more successful than Yeoman II they were funded and released in a slightly less organized way, see Charnley, above n 10.

58 Biffen was knighted in the New Year's honours roll in 1925, receiving, as Middleton and Hall had previously, a KBE, part of the new honours system created at the close of the war to recognize contributions to the nation and empire. David Lloyd George, then Prime Minister, wrote an open letter of support for Weaver, published in *The Times* in 1922. 'Letters to the Editor', *The Times*, 16 January, 4c.

Relations between Public and Commercial Plant Breeders

During the rise of publicly funded plant breeding where were the seed houses and nurseries toured by Shull in 1908? It would be fair to say that in the main, despite Weaver's promises to include them in his plans, they were being side-lined. During the establishment of NIAB the commercial seed trade had sought representation on the institute's executive committees. Although they did eventually gain representation, in the end it was through far fewer committee members than they had hoped for.[59] The British equivalent of the American Seed Traders Association (which had been so vocal in campaigns for the PPA), the Agricultural Seed Traders Association (ASTA) found itself largely on the outside of a circle of influence which it had hoped to buy its way into. The ASTA's initial subscription of around £7,000 towards NIAB did not ensure the sort of representation in the Institute that the ASTA had obviously hoped for.

Relations between publicly funded research institutes and the commercial seed trade remained unfriendly for much of the 1920s. In 1919, Carters, now located within a short distance of Bateson's John Innes Horticultural Research Institute, launched a complaint about NIAB's plans to register a trademark. The nub of the firm's complaint was with the inclusion of the word 'national' in NIAB's trademark. Carters had been using the same word in their trademark since the previous year and felt NIAB's use of the word would overstate the importance of NIAB's seeds, implying that they were the only national alternative for seed (Carters operated across the country), or worse, causing people to confuse NIAB's seeds for Carters'. The case rumbled on for several years, and in the end neither NIAB nor Carters used the word national in their trademarks, probably at least in part because, 'The view of the President of the Board of Agriculture and Fisheries that he strongly deprecates any individual firm of English seed merchants obtaining a monopoly of the word "National" in connection with the seeds sold by them has already been communicated to the patents office.'[60] Of course, NIAB never stopped being the *National* Institute and monopolizing the word through its institutional name – so this block fell much more heavily on seed merchants. In contrast to the American picture then, Britain's commercial plant breeders in the 1920s were struggling to even secure the trademarks they wanted from an unsympathetic legislature.[61]

59 See *Lawrence Weaver to Mr Sutton, 18 July 1918*, Correspondence 1916-1918. Archives of the National Institute of Agricultural Botany, and SW Farmer's concerns about the seed trade's influence, *S. W. Farmer to R. E. Prothero, 27 December 1918*, Correspondence 1916-1918. Archives of the National Institute of Agricultural Botany.

60 *National Institute of Agricultural Botany. Registration of the Word National: Counter Statement*, Trade Mark Case. Archives of the National Institute of Agricultural Botany.

61 See also the Wuesthoff's failed attempts to patent plants in mid-1930s Britain, see Llewellyn and Adcock, above n 8, 138.

Conclusions

If we return to the three comparative categories suggested at the start of this chapter: geographic, scientific and social, it should now be obvious that the third category contains the most contrasts between the two contexts. British plant breeders also faced the threat of piracy in expanding markets, in Europe and internationally. Likewise, the advances in science which seem to have been important in America, if primarily sociologically, were just as important in Britain. Looking to the institutional position of star breeders, the relative mixture of commercial and public breeding and the nature of public breeding and distribution undertaken, differences between the two contexts become more marked. In the first decade of the century, few commercial British nurseries practised their own breeding, and when they did it was arguably for show. Over the next two decades, the stars of British plant breeding were almost exclusively situated in publicly-funded contexts and had little sympathy for commercial plant breeders. At just the point when commercial seed firms were thriving in America the opposite was true in Britain where commercial firms were increasingly competing with publicly-funded breeding. These developments made it highly unlikely that anything like the PPA would be on the British legislative table. Ironically, when calls for PPA-like legislation came, it was while Winston Churchill was in his second term as Prime Minister. Churchill, who had done much to aid Lloyd George in shaping publicly-funded plant breeding, was, in 1954, head of a government that sought to encourage commercial breeders, for the good of the nation. If historically, both private and public enterprise has been called on to serve the nation, in Britain in the 1930s, plant breeding was seen as an activity to be conducted by the state, and one too important to the nation to be the subject of individual profit.[62] In this context, patentable subject matter, *contra* Hansen at the 'Copyright' session, was defined instrumentally, as part of a certain configuration of the patent bargain, rather than by reference to gifts of God.

Turning back to the US case, with these historical insights about what was happening in Britain in mind, history is an integral, if very much contested feature of US legislation. Seventh Circuit Federal Judge, Richard Posner, in a recent review of Supreme Court Justice Antonin Scalia's book-length defence of textual orginalism, summarized the appeal of such a view of legislative history for Scalia, one in which historical context counts for nothing, in the following words: 'Judicial

62 In the 1950s when Churchill and his Conservative government were trying to stimulate private enterprise the Engholm Commission on Plant Breeders' Rights, they instituted, concluded that IP rights of some sort, for breeders, were exactly the right tool for this job. They were unconcerned about questions of specification or the particular type of plant breeding method used and untroubled by the notion that plant breeders were inventors and held no qualms about gifts of God. See 'Plant Breeders' Rights'. 1960. [Cmnd. 1092] *Report of the Committee on Transactions in Seeds 1959-60*. London, Her Majesty's Stationary Office. See also Llewellyn and Adcock, above n 8, 141-142.

historiography rarely dispels ambiguity. Judges are not competent historians. Even real historiography is frequently indeterminate, as real historians acknowledge.'[63]

Yet Scalia's brand of a-historical legislative interpretation, as Posner demonstrates with much glee, does involve making historical judgment. By Posner's reckoning, Scalia employs historical thinking just as incompetently as any other judge, textual originalist or not. The only difference is that Scalia, on this view, dis-indigenously hides his history from sight.

In the recent Supreme Court appeal decision, *Bilski v. Kappos*,[64] and in the petition for a writ of certiorari that led to the hearing, there were 11 references to economic historian Christine MacLeod's *Inventing the Industrial Revolution: The English Patent System 1660-1800*.[65] The details of this case centred on the patentability of a business method as a process patent, and the appropriate way to define a process (hinging in turn on whether or how to apply the machine or transformation test – MORT). In the decision and petition MacLeod's work was brought forward as proof of the original intention of the British patent system as to the patentability of processes. Another example of history being used by legislators, more relevant to plant breeding but less well known, can be found in the final decision of the US Patent and Trademrk Office (USPTO) Appeal Board *In Re Beineke*. After a long series of appeals by Walter Beineke over a discovered oak variety which Beineke was trying to protect with a plant patent, the USPTO's rejection of his patent under section 161 (that varieties discovered in an uncultivated state could not be protected) was affirmed by the Board of Appeals. The Board's decision was premised on a lengthy historical reconstruction of the sections' genesis. The board concluded, 'This history demonstrates that the 1930 Act was not meant to include plants discovered by chance by plant explorers and the like.'[66]

When MacLeod published *Inventing* in 1988 it was a trail-blazing work into an obscure area of economic history. If MacLeod, in 1988, might be forgiven for not having anticipated the ends to which her work would be put, the same cannot be said to apply to writers in this area today. Nowadays, *inventing* is a standard introduction for a well-nourished field of inquiry; IP studies (for want of a better expression to gloss together the various legal studies, economic history, history

63 Posner on Scalia: Posner, R. 2012. The Incoherence of Antonin Scalia, *The New Republic*. [Online]. Available at http://www.tnr.com/article/magazine/books-and-arts/106441/scalia-garner-reading-the-law-textual-originalism?page=0,1 [accessed: 12 November 2012]. Posner is in fact paraphrasing an argument for textual originalism given by Scalia. However, Posner makes abundantly clear in the rest of the article that textual originalism (at least Scalia's version) is equally guilty of the sins of bad history as loose constructionism.

64 *Bilski v. Kappos*, 561 U.S. 8 (2010).

65 MacLeod, C. 1988. *Inventing the Industrial Revolution: The English Patent System, 1660-1800*. Cambridge: Cambridge University Press.

66 *In re Beineke* 690 F.3d 1344, 1352 (C.A.Fed., 2012).

of science and technology, science and technology studies and innovation studies work focused on IP and its history). IP Studies is an area of academia where historical reconstructions are used to further normative decisions about patent law and policy. An even more striking contemporary example of history and law interacting in this way can be seen in Dibner Family Professor of the History & Philosophy of Science & Technology, Myles Jackson's work with the Ameican Civil Liberties Union on the BRCA 1 and 2 and CCR5 gene patents. Finally, Boldrin and Levine's outrageously polemic *Against Intellectual Monopoly*, shows that some academics are now completing the circle themselves and using history to make their own arguments about legislative change.[67] In Boldrin and Levine's case, the argument is derived from their particular brand of neo liberalism, and the view that IP is an unfair monopoly on what should be a freer market of ideas.

Judges and academics are not alone in using history; the legislature has also created conditions favourable to the production of historical analyses of IP. The British and American governments and the EU parliament have been explicit in promoting IP as a driver of the new information economy.[68] Research funding has largely followed that ideology and IP Studies scholars have followed that funding. The turn to innovation in academia, whether academics like it or not, is part of a political trend towards increasing use of IP.

In this context of history–law–politics–policy interaction, what lessons for contemporary debates can be gleaned from Britain's absence of PPA-like legislation? While it might be tempting to read this history as evidence that we ought to remove patents from plants, as Boldrin and Levine at their most strident do, this reading elides many of the interesting differences between then and now. Indeed, there are no direct equivalences, and no direct lessons, or few, which can speak across the contingent differences between past and present. History has greater value, in a smaller, indirect sense, as an experimental plot – a place where we can see different ideas and configurations, than those currently in sway, interacting with each other.

Plant breeders meeting at the RHS's conference were definitely concerned with their IP. They might not have known exactly what they wanted, or shown much agreement about the subject but using copyright for the title of a session that they (or Wilks) felt important enough to record in the conference proceedings

---

67 Boldrin and Levine, above n 9.

68 For the UK situation see, Hargreaves, I. 2011. *Digital Opportunity: A Review of Intellectual Property and Growth*. [Online]. Available at: http://www.ipo.gov.uk/ipreview-finalreport.pdf [accessed: 7 February 2013]. For the US see, USPTO and Economics and Statistics Administration. 2012. Intellectual Property and the U.S. Economy: Industries in Focus. [Online]. Available at: http://www.esa.doc.gov/sites/default/files/reports/documents/ipandtheuseconomyindustriesinfocus.pdf [accessed: 7 February 2013]. Finally, for the EU's IP policy see: European Commission. 2011. Industrial Innovation: Intellectual Property. [Online]. Available at: http://ec.europa.eu/enterprise/policies/innovation/policy/intellectual-property/index_en.htm [accessed: 7 February 2013].

shows that IP was certainly on their agenda. Reconstructing such debates does much to dispel the notion that IP is something defined by reference strictly to the legislative enshrinement of IP in patents, copyright or trademarks. In Paul and his colleagues' discussions, although copyright was an entry point, we can see a much broader but no less valid concept of IP being questioned.

Discussions by plant breeders about rewards and how to balance these with open circulation of materials have been a key feature of the current shift from UPOV to gene patents. However, the original form of UPOV probably owes a great deal more to the sorts of arguments advanced by Biffen and his colleagues than those made by Paul and his peers. A generation after Biffen, Britain figured large in the 1960s debates about farmers' and breeders' exemptions in UPOV, with British delegates arguing against Association Internationale des Selectionneurs pour la Protection des Obtentions (ASSINSEL) delegates for more extensive exemptions.[69] What is more, the idea of serving the purpose of a higher interest (for Biffen: nation and empire) is still with us today, in 'good of the planet' arguments. However, there is a significant difference between Biffen's and modern formulations, where Biffen placed himself – as a university-based public servant – above the use of patents, current thinking stresses that IP is necessary for the good of the planet (specifically to drive a new green economy and the sorts of genetically modified organism (GMO) interventions many are now arguing are necessary to feed a population of seven billion).[70]

History matters, for law, for politics and for policy. Professional historians of IP need to recognize the ends to which their work is likely to be put, especially, as with Myles Jackson, when they are increasingly called upon personally to enter contemporary debates. The historical reconstruction offered here points up two historical lessons for today's debates: (1) IP rights have in the past covered a much broader range of claims than those instantiated in patents, copyright and trademarks.[71] (2) The patent bargain that we see in effect today, in which the rights of inventors (often actually corporations) are set at a much higher value than the public good, is a relatively new development, completely at odds with the non-patent bargain offered to plant breeders in Britain in the 1920s.

69 Bonneuil, C. 2011. *Why the Dwarf Genes of the Green Revolution Were Not Patented.* Workshop paper, presented at *Plants, Animals, and Ownership: Innovation and Intellectual Property Protection in Living Organisms since the 18th Century*. Yale University. On file with the author.

70 See, for example, Thompson, P. 2011. *Agro-Technology: A Philosophical Introduction.* Cambridge: Cambridge University Press.

71 Something of this breadth can still be seen in the specious hard and soft IP distinction.

Chapter 7

Changing the Recipe: Food Security and Other Socio-Economic Considerations in Agricultural Biotechnology Regulation

Karinne Ludlow

Consumer demands, and the expectations of society more broadly, that socio-economic considerations (SECs) be included in regulation of food are growing. The application of biotechnology to food plants, already a contested issue, is one area where an increasing role for SECs in regulation is being debated. There are growing calls for SECs, such as food security, to be included in biosafety assessment processes and IP protection of biotechnology. Similarly the sourcing of genetic resources, often used in agricultural biotechnology, is leading to calls for SECs to be dealt with by IP regimes.

In October 2012 the parties to the *Cartagena Protocol on Biodiversity* (CPB) met to address the issue of the inclusion of SECs in biosafety assessment processes.[1] The meeting's decision to establish an Ad Hoc Technical Expert Group (AHTEG) to develop conceptual clarity opens the way to developing guidelines on such processes and can be expected to inform governments of at least some countries on how their domestic policy should be shaped. These moves to include SECs, including the implications of IP, in biosafety assessments are occurring at a time when the cost and complexity of some existing regulatory systems for genetically modified (GM) crops, such as the European Union system, along with the lack of evidence of harm to people or the environment, 'is generating pressure to make regulatory systems more risk-based and to take potential benefits into account'.[2]

After first examining SECs that may arise in the context of agricultural biotechnology, this chapter examines how SECs have been, or are proposed to be, included in international agricultural biotechnology regulation and the IP regimes that protect it. It then considers parallel moves in relation to Australian biotechnology regulation. Finally, it discusses the problems that need to be

1 Sixth Meeting of the Conference of the Parties to the Convention on Biological Diversity (CBD) serving as the meeting of the parties to the Cartagena Protocol on Biosafety (COP/MOP 6), 1-5 October 2012, India.

2 Tait, J. and Barker, G. 2011. Global food security and the governance of modern biotechnologies. *EMBO Reports*, 12(8), 763-768, 766.

addressed for SECs to be usefully included in agricultural biotechnology regulation and makes some suggestions for future progress.

What are the Relevant SECs?

SECs in the context of agricultural biotechnology are a 'broad spectrum of concerns about the actual and potential consequences of biotechnology'.[3] Providing a taxonomy of SECs relevant to agricultural biotechnology though is a difficult task – although there has been much discussion and some work by certain jurisdictions,[4] there is still a lack of information on the content or meaning of relevant SECs. Various bodies are beginning to compile possible lists of SECs such as the Third World Network,[5] Chatham House, UK[6] and the World Resources Institute.[7] For example, the Netherlands Commission on Genetic Modification (COGEM) published a report in 2009 on the potential role of socio-economic assessments as part of biosafety approval processes.[8] The COGEM report suggested that the criteria most relevant to Europe, but which have not yet been included in an assessment framework, could be summarized down to three themes – (1) benefit to society, (2) economics and prosperity, and (3) cultural heritage.[9]

---

3 La Vina, A. and Fransen, L. 2004. *Integrating Socio-Economic Considerations into Biosafety Decisions: The Challenge for Asia*, 3. [Online]. Available at: http://fbae.org/2009/FBAE/website/images/pdf/imporatant-publication/integrating_socio_econmic_needs_in_biosafety.pdf [accessed: 1 February 2013].

4 See, for example, European Commission. 2011. *Final Report from the Commission to the European Parliament and the Council on socio-economic implications of GMO cultivation on the basis of Member States contributions, as requested by the Conclusions of the Environmental Council of Dec 2008*. [Online]. Available at: http://ec.europa.eu/food/food/biotechnology/index_en_htm [accessed: 3 January 2013]; Netherlands Commission on Genetic Modification (COGEM). 2009. *Socio-economic aspects of GMOs. Building Blocks for an EU Sustainability Assessment of Genetically Modified Crops*. [Online]. Available at: http://www.cogem.net/main-adviesdetail-signaleringEN.aspx?pageid=54&loc=2&version=&mode=&id=517 [accessed: 1 February 2013].

5 Ling, C.Y. 2008. *Assessing the socio-economic, cultural and ethical impacts of GMOs: Third World Network Briefings for MOP 4#3*. [Online]. Available at: bch.cbd.int/cms/ui/forums/attachment.aspx?id=301 [accessed: 1 February 2013].

6 Jarvis, A. 2009. *A Way Forward for GMO Cultivation in Europe?* [Online]. Available at: http://www.chathamhouse.org/publications/papers/view/109176 [accessed: 1 February 2013].

7 Fransen, L., La Vina, A., Dayrit, F., Gatlabayan, L., Santosa, D.A. and Adiwibowo, S. 2005. *Integrating Socio-economic Considerations into Biosafety Decisions: The Role of Public Participation*. Washington DC: World Resources Institute.

8 COGEM, above n 4.

9 Ibid., 5.

A more detailed list was compiled through a 2009 survey for the CPB process.[10] The survey was undertaken after the inclusion of SECs in biosafety assessments was discussed at the 2008 CPB's Conference of the Parties/Meeting of the Parties (COP8/MOP8) and it was decided further technical guidance was needed. In 2009, a survey of countries' experiences with SECs in biosafety decision making was conducted through the United Nations Environment Programme's Division of Global Environment Facility Coordination (UNEP DGEF).[11] The meeting of the CPB parties in October 2012 considered this information. The survey identified the five most important SECs as:

- food security;
- health-related impacts;
- coexistence of GM organisms (GMOs);
- impacts on market access;
- compliance with biosafety measures.

The next five SECs were variously drawn from:

- impacts on the conservation and sustainable use of biodiversity;
- economic impacts of changes in pest prevalence;
- macroeconomic impacts;
- farmers' rights;
- IP;[12]
- consumer choice;
- economic impacts of changes in use of pesticides and herbicides;
- indigenous and local communities.[13]

Defining the content of any particular SEC is difficult and dependent on context. The SECs are tightly interwoven, and of course, different countries and groups within countries (such as consumers, producers and industry) interpret

10 Conference of the Parties Serving as the Meeting of the Parties to the Cartagena Protocol on Biosafety. 2010. *Summary Report on the Survey on the Application of and Experience in the Use of Socio-Economic Considerations in Decision-Making on Living Modified Organisms*, UNEP/CBD/BS/COP-MOP/5/INF/10. Montreal: Secretary to the Convention on Biological Diversity.

11 Information about the survey is available at Biosafety Clearing-house. 2009. *Welcome to the Survey on Socio-Economic Considerations*. [Online]. Available at: http://bch.cbd.int/onlineconferences/socioeconomics.shtml [accessed: 1 February 2013].

12 The survey also found that respondents from countries with a decision making system under the CPB in place or under development, considered IP as a more important SEC than respondents from countries without such a system: see UNEP/CBD/BS/COP-MOP/5/INF/10, above n 10, 6.

13 Ibid., 5.

and are affected by any particular SEC in different ways. How a particular SEC is relevant to a specific country and what methodologies are appropriate in an assessment of that will also vary depending on, for example, whether the country is a developed, developing or least developed nation, because this determines the infrastructure and human resources available, and the usual governance patterns used in the nation. A SEC's impact will also differ because of different cultural and religious values, political and economic capacity and forms of agriculture and biotechnology research undertaken in particular countries.

Food security, clearly an important SEC in the context of agricultural biotechnology, is an example of the difficulty of defining the content of a SEC. The United Nations Food and Agriculture Organization (FAO) defines food security as existing 'when all people, at all times, have physical and economic access to sufficient, safe and nutritious food to meet their dietary needs and food preferences for an active and healthy life'.[14] Some interpret food security then as simply meaning access to sufficient food. So, for example, Brookes understands the SEC of food security as part of the ability of households to afford to feed themselves.[15] Others interpret food security as meaning access to the type of food that the consumer wants (preferences), and so for them food security includes choice. Yet others define it with wider dimensions as including conservation of biological diversity so that sufficient food can be ensured and yet others expand it to include the nature of agriculture itself insisting that at the least, the form of agriculture used must be sustainable. For example, Diblin *et al.* say:

> in debating the prospective contribution of GMOs to food security, it is important to consider dimensions of the concept beyond basic food sufficiency: production must be sustainable, increases in production must be distributed equitably, and the food produced must be affordable and consistent with consumers' perceived

---

14 Food and Agriculture Organisation of the United Nations. 1996. *Rome Declaration on World Food Security, World Food Summit.* [Online]. Available at: http://www.fao.org/wfs/index_en.htm [accessed: 30 July 2010]. This definition is also used in Australian policy setting. For example, The Prime Minister's Science, Engineering and Innovation Council. 2010. *Australia and Food Security in a Changing World*, 9. [Online]. Available at: http://www.innovation.gov.au/Science/PMSEIC/Documents/AustraliaandFoodSecurityinaChangingWorld.pdf [accessed: 1 February 2013]; Department of Agriculture, Fisheries and Forestry. 2012. *National Food Plan Green Paper 2012*, 57 [Online]. Available at: http://www.daff.gov.au/__data/assets/pdf_file/0009/2175156/national-food-plan-green-paper-072012.pdf [accessed: 1 February 2013].

15 The four references Brookes describes as relevant to food security are put under the heading 'Impact on household incomes & food security' and all relate to this issue. See Brookes, G. 2009. *Socio-economic Impacts of GM Crop Technology: 'Second Round' Impacts – Briefing Note*, 13. [Online]. Available at: http://ec.europa.eu/food/food/biotechnology/reports_studies/docs/Europabio_contribution_I_en.pdf [accessed: 1 February 2013].

needs. Broader understandings of food security thus include questions about access and utilisation as well as simple availability.[16]

Further, for some the increasing use of GM crops to provide feedstock for alternative forms of energy, such as for biofuels, raises concerns that the use of agricultural land for that purpose puts increasing pressure on not using that land for food, and so poses a food security risk.[17] The Third World Network sees this as a key concern within food security, for the purposes of developing countries, raising the issue of the impact of adoption of GM crops by subsistence farmers who may also supply domestic markets with food, for industrial use rather than use as food.[18] A 2010 PMSEIC Report analysed food security as being composed of five pillars: availability, accessibility, acceptability, adequacy and stability, which cover all of the issues raised above and more.[19]

Some consider that agricultural biotechnology has much to offer in addressing both domestic and international food security problems. Others argue that GM technology undermines food security. Claims have been made that agricultural biotechnology is relevant to food security only because the commercialization of GM products is legitimatized 'partly by the supposed necessity to "feed the world"'.[20] One point is clear: 'The contribution of agricultural biotechnology to food security, whether as benefactor or threat, thus remains as hotly contested as ever.'[21]

Compiling a list of SECs relevant to IP in agricultural biotechnology, particularly patents over GM traits and processes, is similarly difficult, with intertwining themes and issues. But in the context of food security, after describing the risk of contamination of non-GM crops by GM crops as one of the main threats to Australian food security posed by GM crops, Diblin et al. suggest a further threat is 'the expansion in control of seeds by agricultural biotechnology corporations'.[22] The same concern is raised in the context of developing countries by the Third World Network.[23] Diblin et al. conclude that there is 'increasing evidence of influence of giant seed and biotechnology corporations on agricultural and food policy. Corporate control of seeds has been widely viewed by critics as

16 Dibden, J., Gibbs, D. and Cocklin, C. 2013. Framing GM crops as a food security solution. *Journal of Rural Studies*, 29, 59-70, 59-60.

17 Austin, M. and Kirkwood, L. 2012. Queensland's Strategic Cropping Land Act 2011 and extractive industries: Let's look beyond the borders. *Australian Environment Review*, 27(3), 75-79, 75.

18 Ling, above n 5, 3.

19 Prime Minister's Science, Engineering and Innovation Council, above n 14, 9-10.

20 Dibden et al., above n 16, 60.

21 Dibden et al., above n 16, 69. This article also contains an interesting analysis of use of risks and solutions to food security in debates about agricultural biotechnology.

22 Dibden et al., above n 16, 67.

23 Ling, above n 5.

threatening food security and food sovereignty – i.e. control of domestic food production.'[24] The diversion of public R&D funding into the technology, the loss of the opportunity to pursue other alternatives to the technology and the capture of the outcomes of public funding of biotechnology research by private companies are all also debated as SECs relevant to IP in agricultural biotechnology. The Third World Network raise the hampering of the free flow of information, knowledge and genetic materials that is needed for R&D in public universities as a further SEC relevant to IP in agricultural biotechnology.[25]

Regardless of the exact content of 'food security' as a SEC, agricultural biotechnology's position on the boundary between agriculture (that is, living things) and food (commodities) is probably better described as being on a fault line in terms of the policy issues it raises. This is clearly demonstrated by the failure of the ambitious intergovernmental scientific panel announced in 2002 as a result of the World Summit on Sustainable Development. That panel, the International Assessment of Agricultural Knowledge, Science and Technology for Development (IAASTD), was launched in 2004 with a mandate to 'forecast and propose responses to the agricultural and food security challenges facing the world, from [2004] until 2050'.[26] It was to assess the state of the field using established rules of scientific enquiry. The field was itself divided into eight themes, some of which are the SECs identified above as relevant to agricultural biotechnology (e.g. human health, trade and markets, traditional and local knowledge and community-based innovations, and women in agriculture), although biotechnology was itself treated as a separate theme.[27] For reasons still in debate, the assessment failed. In relation to biotechnology itself, other than acknowledging the debates over the issue, the key document produced by the assessment made no progress on making an assessment or even explaining how to assess these issues. It only calls for 'new kinds of support for the public to critically engage in assessments of the technical, social, political, cultural, gender, legal, environmental, and economic impacts of modern biotechnology'.[28]

24 Dibden et al., above n 16, 68.

25 Ling, above n 5.

26 Edwards, L. 2012. Food fight: International Assessment of Agricultural Knowledge, Science and Technology for Development. *AgBioForum*, 15(1), 70-76, 72.

27 Ibid., 74.

28 Ibid., 75 quoting International Assessment of Agricultural Knowledge, Science and Technology for Development. 2009. *Agriculture at a Crossroads: International Assessment of Agricultural Knowledge, Science and Technology for Development (Synthesis Report)*. [Online]. Available at: http://www.agassessment.org/reports/IAASTD/EN/Agriculture%20at%20a%20Crossroads_Synthesis%20Report%20(English).pdf [accessed: 1 February 2013].

SECS and International Agricultural Biotechnology Regulation

Within the context of the *Convention on Biological Diversity* (CBD), the CPB is an implementing agreement addressing biosafety. The meaning of 'biosafety' is controversial, but for the purposes of this chapter is the safe use of biotechnology and its products, in particular those precautionary practices intended to ensure the safe transfer, handling, use and disposal of living modified organisms (LMOs) and their products.[29] The CPB governs the transboundary movements of LMOs and establishes the rights of recipient countries to be notified of and to approve or reject domestic import or production of LMOs. The CPB uses the term 'LMO' to restrict its application to living organisms and thus excludes processed GM commodities from its scope. It is important to note here that GM food, one of the most controversial aspects of agricultural biotechnology, is therefore not included within the scope of the CPB. Conversely, for many countries, domestic regulation, such as Australian regulation of food and gene technology, applies to GM food or GMOs rather than only to LMOs.

Decisions under the CPB regarding import or production are to be on the basis of a biosafety assessment. Controversially, article 26.1 of the CPB provides that SECs may be included within the domestic measures implementing the Protocol:

> 1. The Parties, in reaching a decision on import under this Protocol or under its domestic measures implementing the Protocol, may take into account, consistent with their international obligations, socio-economic considerations arising from the impact of living modified organisms on the conservation and sustainable use of biological diversity, especially with regard to the value of biological diversity to indigenous and local communities.
>
> 2. The Parties are encouraged to cooperate on research and information exchange on any socio-economic impacts of living modified organisms, especially on indigenous and local communities.

There is debate about the proper interpretation of this article, such as whether an SEC must impact on indigenous and local communities to be relevant.[30] Nevertheless, many developing countries in particular have expanded their interpretation of

29 Convention on Biological Diversity, *The Cartagena Protocol, Resources, Frequently Asked Questions, FAQs 2 and 9*. [Online]. Available at: bch.cbd.int/protocol/ [accessed: 1 February 2013].

30 Jaffe, G. 2005. Implementing the Cartagena Biosafety Protocol through national biosafety regulatory systems: An analysis of key unresolved issues. *Journal of Public Affairs*, 5(3-4), 299-311.

article 26 to include a broad range of SECs.[31] Such countries include Indonesia,[32] South Africa,[33] The Philippines[34] and (a developed country) Norway.[35] Indonesia's legislation, for example, provides: 'This government regulation is based on the precautionary approach to contribute to ensuring environmental safety, food safety and or animal feed safety based on an accurate scientific method by considering religious, ethic, socio-cultural and esthetic [sic] norms.'[36]

The Philippines regulation says: 'The socio-economic, ethical and cultural benefits and risks, of modern biotechnology to the Philippines and its citizens, and in particular on small farmers, indigenous peoples, women, small and medium enterprises and the domestic scientific community, shall be taken into account in implementing the [National Biosafety Framework].'[37]

In contrast the Norwegian Act says that: 'In deciding whether or not to grant the application, significant emphasis shall also be placed on whether the deliberate release represents a benefit to the community and a contribution to sustainable development.'[38]

Other countries, particularly the agricultural export driven developed countries of Australia, the United States and Canada (which are not CPB parties), consider that biosafety assessments should be conducted on the basis of a science-based risk assessment only whereby scientific data is gathered and standard methodologies are used to determine whether the risk is one the jurisdiction is willing to take in allowing the deliberate release of LMOs into the environment. The considerations that are assessed in this manner are generally limited to risks to human health or

31 Falck-Zepeda, J. and Zambrano, P. 2011. Socio-economic considerations in biosafety and biotechnology decision making: The Cartagena Protocol and National Biosafety Frameworks. *Review of Policy Research*, 28(2), 171-195.

32 Secretariat of the Convention on Biological Diversity. 2008. *Regulation of the Government of the Republic Indonesia Number 21 Year 2005 on Biosafety of Genetically Engineered Product*, Art 3 (English translation). [Online]. Available at: http://bch.cbd.int/database/record.shtml?documentid=11576 [accessed: 1 February 2013].

33 Secretariat of the Convention on Biological Diversity. 2012. *Socio-economic considerations*. [Online]. Available at http://bch.cbd.int/protocol/outreach/factsheets/factsheet_socioeconomic_en.pdf [accessed: 1 February 2013].

34 Executive Order No. 514, National Biosafety Framework of the Philippines, Principle 2.5 Socio-economic, Cultural, and Ethical Considerations.

35 *Norwegian Gene Technology Act* (No 38/1993). See further, Myhr, A.I. and Rosendal, G.K. 2009. *GMO assessment in Norway as compared to EU procedures: societal utility and sustainable development. Trondheim, Norway: The Directorate for Nature Management*. [Online]. Available at: www.dirnat.no/attachment.ap?id=104 [accessed: 1 February 2013].

36 *Regulation of the Government of the Republic Indonesia Number 21 Year 2005 on Biosafety of Genetically Engineered Product*, above n 32, Art 3.

37 Executive Order No. 514, National Biosafety Framework of the Philippines, Principle 2.5 Socio-economic, Cultural, and Ethical Considerations.

38 *Norwegian Gene Technology Act* (No 38/1993) s. 10.

safety or environmental risks. Any risks or benefits to SECs are excluded as is the use of social factors in risk evaluation.

In parallel with the developments under the CPB auspices, the European Union has also undertaken work on the SECs in GMO cultivation. The European Directive 2001/18/EC[39] on the deliberate release of GMOs into the environment states that 'Member States may take into consideration ethical aspects when GMOs are released'.[40] This allows for different national standards on ethical judgment but there is no common definition or understanding of 'ethical judgment'. This means different interpretations are being used in various European Union members. For example, Finnish law focuses only on ecological and health risks but Swedish law considers physical and societal effects, and the Swedish Office of the Auditor General reports that this is taken to include societal utility.[41]

The Directive also requires the European Commission to provide a report after three years of a GMO release on the Directive's implementation, including an assessment of the socio-economic implications (defined to mean the advantages and disadvantages[42]) of deliberate releases and placing on the markets of GMOs. Although a report was submitted in 2004, the Commission raised the lack of experience with the issue as a hurdle in making an assessment. Therefore in 2008 the European Council invited further work on the issue by the European Commission and the Member States. The report was provided to the Council in 2011. This and its accompanying working paper[43] provide a useful snapshot of what the Member Countries are doing in respect of the use and implications of GMOs but do not provide any detail on how SECs are or could be included in regulatory frameworks.

There is clearly the potential for conflict with a nation's international obligations under the World Trade Organization (WTO) agreements. Article 26.1 of the CPB recognizes this when it states that the inclusion of SECs must be done consistently with the country's international obligations. Two WTO Agreements are of particular importance to agricultural biotechnology generally, the *Agreement on the Application of Sanitary and Phytosanitary Measures* (SPS Agreement) and

39 European Parliament and European Council. 2001. *Directive 2001/18/EC of the European Parliament and of the Council of 12 March 2001 on the Deliberate Release into the Environment of Genetically Modified Organisms and Repealing Council Directive 90/220/EEC*. Official Journal of the European Communities, 17.4.2001, L 106/1-37, 1.

40 Ibid., Preamble (Point 9).

41 Myhr and Rosendal, above n 35, 12-13.

42 European Parliament and European Council, above n 39, Recital 62.

43 European Commission. 2011. *Commission Staff Working Paper accompanying document to the Report from the Commission to the European Parliament and the Council on socio-economic implications of GMO cultivation on the basis of Member States contributions, as requested by the Conclusions of the Environmental Council of December 2008*. [Online]. Available at: http://ec.europa.eu/food/plant/gmo/reports_studies/docs/swp_socio_economic_report_gmo_en.pdf [accessed: 1 February 2013].

the *Agreement on Technical Barriers to Trade* (TBT Agreement), and a third, *Trade-related Aspects of Intellectual Property Rights* (TRIPS), is particularly relevant to IP.

Inevitably, different countries are variably susceptible to different SECs. Their ability to respond to those considerations though is constrained by their obligations to maintain international free trade. For example, the SPS Agreement concerns food safety and the regulation of animal and plant health. It recognizes that countries have the right to take sanitary and phytosanitary measures but requires that such measures be applied only to the extent necessary to protect human, animal or plant life or health so as to minimize trade barrier impacts and should not arbitrarily or unjustifiably discriminate between members of the WTO where identical or similar conditions prevail. Members of the WTO are also obliged to ensure that as far as possible their quarantine measures conform to international standards, guidelines and recommendations and are based on an assessment of the risk to human, animal and plant life or health, 'taking into account' risk assessment techniques developed by relevant international organizations.[44]

International IP Regulation and Agricultural Biotechnology

A second and more recent agreement under the CBD, the Nagoya Protocol,[45] is further evidence of the growing creep of SECs into agricultural biotechnology regulation. Pursuant to the Nagoya Protocol, countries granting access to bioresources must, amongst other things, ensure domestic procedures require prior informed consent on mutually agreed terms by any indigenous and local groups that hold relevant genetic resources or traditional knowledge,[46] ensure proper records are kept of the origin and use of genetic resources and benefits obtained from them and respect customs, traditions, values and customary practices of

---

44 The SPS Agreement spells out procedures and criteria for assessing risk and determining appropriate levels of health protection. For more on the SPS Agreement in the context of agricultural biotechnology, see Peel, J. 2007. A GMO by any other name ... might be an SPS risk!: Implications of expanding the scope of the WTO Sanitary and Phytosanitary Measures Agreement. *European Journal of International Law*, 17(5), 1009-1031.

45 The Nagoya Protocol on Access to Genetic Resources and the Fair and Equitable Sharing of Benefits Arising from their Utilization: Conference of the Parties to the Convention on Biological Diversity. 2011. *Report of the Tenth Meeting of the Conference of the Parties to the Convention on Biological Diversity*, UNEP/CBD/COP/10/27. Montreal: Secretariat to the Convention on Biological Diversity, [103] and Annex (Decision X/1, Annex 1 (89-109) (Nagoya Protocol).

46 Ibid., Art 7.

indigenous and local communities.[47] Countries are also to consider the importance of genetic resources for food and agriculture for food security.[48]

An important ongoing debate arising out of these requirements is whether patent applicants should have to disclose the use of traditional knowledge or genetic resources in their invention[49] so that these obligations can be met. The Protocol is deliberately vague on whether this is required, requiring instead that countries take 'appropriate, effective and proportionate measures to address non-compliance'. There are other options that can be used to ensure those accessing bioresources comply with the Protocol's obligations, but the fact that the issue is being investigated, even in Australia, shows that IP regimes are not outside the growing push for SECs to be part of the agricultural biotechnology regulatory regime.

It should be noted that the Nagoya Protocol does not apply to genetic resources already covered by specialized access and benefit-sharing agreements such as the *International Treaty on Plant Genetic Resources for Food and Agriculture* (*Plant Treaty*).[50] The Treaty applies to 64 food and forage plants which are considered of the greatest importance for food and agriculture. Whilst many GM crops would fall within this scope, the plants or other living things from which genes are originally sourced may not. In that case, the Nagoya Protocol would be relevant.

In the WTO context, TRIPS provides that a minimum threshold of IP protection must be provided to all technologies, but it does allow for countries to provide limited exceptions to patent protection.[51] Some countries have used this provision to regulate IP in agricultural biotechnology differently to IP in other technologies. They and others may also now seek to 'assess' the fact that there are IP in GMOs in their decision-making processes regarding agricultural biosafety and provision of access to bioresources. In Australia attempts to exclude GMOs from patentability have been unsuccessful and so there is no difference in protection offered to agricultural biotechnology.[52] As discussed below, it also seems that current governmental policy is not to use the existence of IP in its decision making regarding mandatory food labelling. However, it remains to be seen whether

---

47 Ibid., Art 12.

48 Ibid., Art 8(c).

49 This issue is also now being pursued at the World Trade Organization.

50 Nagoya Protocol, above n 45, Art 4.4.

51 For example, the EU's ordre public exception and see Decision of Examining Division of the EPO *In re President and Fellows of Harvard College* (1993) 24 IIC 103.

52 Bills proposing the inclusion of a further exception to patentable inventions, proposed by Senator Coulter in 1990 (which would have excluded GMOs and their progeny, amongst other things) and Senator Stott Despoja in 1996 and 2001 (which would exclude naturally occurring genes or gene sequences or descriptions of the base sequence of a naturally occurring gene or a naturally occurring gene sequence) have been rejected by the Senate.

operational processes and consumer demands will lead to requirements that such information be provided in either the food labelling or bioprospecting sectors.

Australian Agricultural Biotechnology and IP Regulation

There is growing awareness of the issue of food security in the Australian domestic context with a Senate inquiry into Australia's food production in 2008-10,[53] the Prime Minister's Science, Engineering and Innovation Council (PMSEIC) Expert Working Group 2010 Report on food security which also recommended the establishment of an Australian Food Security Agency,[54] and a July 2012 Green Paper on a National Food Plan that is also investigating the introduction of a body to manage food policy.[55] The growing importance of social values in Australian food purchasing decisions has also been noted.[56]

The Australian IP regimes are also no stranger to demands that SECs be part of the biotechnology regulatory structure, although to date the most vocal calls for changes to Australian patent law on biotechnology have concerned only human biotechnology. For example, the 2003-4 Australian Law Reform Commission inquiry was limited to IP over genetic materials and genetic and related technologies relevant to healthcare.[57] Similarly, the 2008-10 Senate inquiry into gene patents was limited to human and microbial gene patents, and not plant or animal (agricultural) genes.[58] The inquiry was also concerned only with the

53 Senate Select Committee on Agricultural and Related Industries. 2010. *Food Production in Australia: Final Report*. [Online]. Available at: http://www.aph.gov.au/Parliamentary_Business/Committees/Senate_Committees?url=agric_ctte/completed_inquiries/2008-10/food_production/report/index.htm [accessed: 1 February 2013].

54 Prime Minister's Science, Engineering and Innovation Council 2010. Above n 14, Recommendation 1 (63-65).

55 Department of Agriculture, Fisheries and Forestry. 2012. *National Food Plan Green Paper*. [Online]. Available at: http://www.daff.gov.au/__data/assets/pdf_file/0009/2175156/national-food-plan-green-paper-072012.pdf [accessed: 1 February 2013].

56 Department of Agriculture, Fisheries and Forestry. 2011. *Issues Paper to Inform Development of a National Food Plan*, 27 (section 4.3.1). [Online]. Available at: http://www.daff.gov.au/__data/assets/pdf_file/0009/1926315/nfp_-_final.pdf [accessed: 1 February 2013].

57 Australian Law Reform Commission. 2004. *Genes and Ingenuity: Gene Patenting and Human Health*, ALRC Report 99. Sydney: Australian Law Reform Commission.

58 However, a Bill introduced by Senator Xenaphon two days before the Senate report was released was not so limited: see *Patent Amendment (Human Genes and Biological Material) Bill 2010*. It potentially would, if passed, have applied to all 'biological materials' (s 4) regardless of their source (although the name of the Bill may show this was unintended). The Bill was not passed after being referred to Senate Legal and Constitutional Affairs Legislation Committee that reported 21 September 2011 and recommended the Bill not be passed.

impacts of such genes on healthcare, medical research and health and wellbeing of Australians, rather than agriculture or food.[59]

Two national regulators play a lead role in Australian regulation of agricultural biotechnology – the food regulator (Food Standards Australia New Zealand (FSANZ)) and a regulator created specifically to regulate gene technology in Australia (Gene Technology Regulator (GTR)).

Although the Australian approach to GM food regulation has largely been on the basis of food safety, the national *Australia and New Zealand Food Standards Code* contains mandatory labelling requirements for many GM foods and it has never been entirely clear whether this has been, at least in part, in response to consumer demands for information to enable them to respond to SECs. This has become of particular importance as newer technologies, such as nanotechnology, begin to be applied to food. Some guidance on the objective of labelling requirements and whether SECs are important in deciding whether to introduce labelling requirements can be gained from the 2011 *Labelling Logic* Report.

The *Labelling Logic* report was produced by an independent review of food labelling commissioned by the Australian and New Zealand Food Regulation Ministerial Council and saw ethical and consumer values such as environmental sustainability as low risk issues that in most cases demanded labelling regulation only through self-regulation or co-regulation. Environmental concerns are described in the Report as a 'category of consumer demands which cannot easily be linked to either short-term or long-term health risks. These demands arise from consumers' perceptions of the world, their ethical views and their personal values. Information demands reflecting consumer values include country-of-origin labelling (CoOL), environmental issues, animal welfare and methods of production',[60] and obviously include many SECs relevant to agricultural biotechnology. In contrast, food safety was seen as high risk, demanding mandatory labelling regulation through government intervention.[61] The review recommended 'a framework that will guide decision making on food labelling matters using a three tiered hierarchy consisting

59 A Dutch study has found that access to genetic variation is crucial to further innovation in plant breeding: Louwaars, N., Dons, H., van Overwalle, G., Raven, H., Arundel, A., Eaton, D. and Nelis, A. 2009. *Breeding Business. The future of plant breeding in the light of developments in patent rights and plant breeders' rights*. [Online]. Available at: http://documents.plant.wur.nl/cgn/literature/reports/BreedingBusiness.pdf [accessed: 1 February 2013].

60 Blewett, N., Goddard, N., Pettigrew, S., Reynolds, C. and Yeatman, H. 2011. *Labelling Logic: Review of Food Labelling Law and Policy*, Final Report, [2.6]. [Online]. Available at: http://www.foodlabellingreview.gov.au/internet/foodlabelling/publishing.nsf/content/48C0548D80E715BCCA257825001E5DC0/$File/Labelling%20Logic_2011.pdf [accessed: 1 February 2013].

61 Ibid., 86.

of food safety, followed by preventative health and consumer values issues'.[62] However, the Report placed GM labelling at a higher place in the hierarchy than it would be if only consumer value issues were raised and recommended that GM food have mandatory labelling. This was because of concerns about long-term health issues[63] and not because of any acute dangers.

In its response to the *Labelling Logic* Report, the COAG Legislative and Governance Forum on Food Regulation (which replaced the Ministerial Council in 2011) agreed to the suggested three tiered hierarchy for labelling decisions. However, although the Forum agreed that GM food should be labelled, it did not commit to treating all new technologies as automatically requiring mandatory labelling – instead stating that any new technology would be 'considered on a case by case basis'.[64] The *Labelling Logic* Report had recommended mandatory labelling of new technologies for 30 years after their first introduction to the market. But the Forum noted 'that mandatory labelling would place a significant cost burden on industry, may be trade-restrictive, and may inhibit investment in new technologies'.[65] The Forum did though undertake 'to develop a conceptual framework and issues hierarchy that will include a Ministerial Policy Guideline for the case-by-case consideration of regulatory (i.e. labelling) and non-regulatory measures applying to food produced using a new technology requiring pre-market safety assessment'.[66] That Policy Guideline is to highlight the importance of consumer values issues as part of the consideration. In regards to Australia's international obligations, the Forum noted that the proposal for mandatory labelling for new technologies constitutes a technical regulation under the WTO TBT Agreement and that a case-by-case approach is needed to ensure Australian food regulation is consistent with its obligations under the TBT Agreement.[67]

SECs including food security are also outside the scope of the gene technology regulatory framework. Whilst it was originally envisaged that social and ethical issues would also be considered by the GTR when making decisions on gene technology applications, and public consultation during the scheme's creation

---

62 Department of Health and Ageing, *Legislative and Governance Forum on Food Regulation, Communiqué*. [Online]. Available at: http://www.health.gov.au/internet/main/publishing.nsf/Content/4070FF99C785D421CA257961000F0C51/$File/Dept091211.pdf [accessed: 1 February 2013].

63 Blewett et al., above n 60, [2.5].

64 Legislative and Governance Forum on Food Regulation. 2011. *Response to the Recommendations of Labelling Logic: Review of Food Labelling Law and Policy*, 12. [Online]. Available at: http://www.foodlabellingreview.gov.au/internet/foodlabelling/publishing.nsf/content/ADC308D3982EBB24CA2576D20078EB41/$File/FoFR%20response%20to%20the%20Food%20Labelling%20Law%20and%20Policy%20Review%209%20December%202011.pdf [accessed: 1 February 2013].

65 Ibid., 34.

66 Ibid.

67 Ibid., 34-35.

showed the Australian public considered ethical issues equally important to safety and environmental concerns, the final scheme does not address these matters. A 2006 independent review of the scheme concluded that the existing scope of the gene technology regulatory framework should be maintained. Nevertheless, since the establishment of the national scheme there have been calls from some States for SECs to be added to the GTR's risk assessment process. The limited scope of the GTR's assessment process and the proposed commercial release of GM canola in 2003 were important in the decision of many States to introduce their own legislation dealing with SECs. These created moratoria on certain GMO releases in all but two States, the purpose of which was to preserve the identity of non-GMOs for marketing purposes and provide time to consider the socio-economic ramifications of GMO releases. That legislation is still in force, but the moratoria have expired in all States except South Australia and Tasmania, where the ban is likely to continue until at least 2014.[68]

SECs in Decision Making

The inclusion of SECs in regulatory frameworks for GMO assessment (in the case of biosafety or labelling decisions) or ensuring compliance with IP regimes (in the case of access to bioresources) raises many issues in terms of design. Some of the difficulties in even identifying SECs generally have been described above. If SECs are defined broadly enough to be regarded as shared by most in the community, they are likely to become ambiguous.[69] Further, if there is not a sufficient link between broad values and individual decisions, the values will effectively be excluded from the decision-making process.[70] In addition to identifying and defining the SECs to be considered and the issues raised within any particular SEC, many other issues must be addressed to create a functional assessment system that does not impose an unnecessary hurdle to the introduction of GM crops.

For example, it must be decided who should do the assessment, who should have input into the process and at what stage of the introduction of the GMO should the assessment should be done. SEC assessment could be done ex-ante (before) or post-ante (after) the GMO is released into the environment. The European Union report referred to above was compiled because the European Commission requires a report on the SECs of GM crops every three years *after* introduction, although what is meant by SEC and how the information is to be used in the regulatory

68 In Tasmania, all dealings with GMOs, including research, are expressly prohibited unless licensed by the Gene Technology Regulator *and* authorized under State legislation. The South Australian ban applies only to cultivation of GM food crops, with food crop being defined to include crops intended for animal consumption.

69 Hope, J. 2001. New Zealand Royal Commission on Genetic Modification. *Environmental and Planning Law Journal*, 18(5), 441-444, 444.

70 Ibid.

process is unclear. But most regulatory frameworks that already require SEC assessment undertake it ex-ante.[71]

A further issue that must be dealt with by any regulatory framework incorporating an SEC assessment process is how the implications of particular SECs are to be measured, whether only negative or also positive implications should be taken into account, and whether SECs should be assessed at a local, national or international scale. For example, as noted above, SECs are a central part of the Norwegian impact assessment prior to the marketing of LMOs. However, whilst the assessment of sustainability issues (ecological, economic and social) is to be performed on a global scale, implying that considerations should also focus on the country producing the LMO in question, assessment of the benefits to society is to focus primarily on 'effects in Norway, such as whether the LMO helps solving [*sic*] a social problem, create employment opportunities or tends to create problems for existing production that otherwise should be maintained'.[72] In contrast, the South African legislation allows the decision maker to consider SECs that the introduction of a GMO may have only 'on a community living in the vicinity of such introduction'.[73]

In most cases various SECs will be affected differently both in the size of the impact and whether that impact could be said to be a benefit or not for the nation concerned. It therefore also needs to be decided how competing SECs will be balanced against each other and what their order of priority is. If this is not done, there will be no guidance as to how any conflict should be resolved.[74] For example, the New Zealand GM regulatory scheme requires adverse effects to be weighed against beneficial effects on a case-by-case basis.[75] 'Effects' includes potential, probable and future effects. The scheme also provides for social and cultural as well as scientific and economic values to be considered and makes input by Maoris and the consideration of their spiritual beliefs when making decisions mandatory. In considering an application to undertake research into GM cows in which a gene originally sourced by copying a human gene was to be inserted into the cows'

---

71 Falck-Zepeda, J. 2009. Socio-economic considerations, Article 26.1 of the Cartagena Protocol on Biosafety: What are the issues and what is at stake? *AgBioForum*, 12(1), 90-107, 91.

72 Biosafety Clearinghouse (Convention on Biological Diversity). 2008. *Second National Report Analyzer, Article 26 – Socio-economic considerations, question 178, Norway*. [Online]. Available at: http://bch.cbd.int/database/reports/results/?searchid=551899 [accessed: 3 February 2013].

73 Falck-Zepeda, above n 71, 92.

74 Hope, above n 69, 444.

75 *Hazardous Substances and New Organisms Act 1996,* although the risk assessment details are generally set out in the Hazardous Substances and New Organisms (Methodology) Order 1998.

DNA, these issues were addressed.[76] A major cultural consideration was that the relevant Maori group believe GM involving different species is contrary to their *Tikanga* (custom) because it interferes with the *whakapapa* (genealogy) as well as the *mauri* (life force or essence) of both species. The Maoris believe that their spiritual guardianship extends to imported species that have had a long presence in New Zealand and could be regarded as valued species and therefore includes cows although these are not native animals. It is also believed that proceeding with GM of any kind, but particularly where genes were moved between species, may result in members of their group suffering adverse health consequences and even death. The New Zealand regulator concluded on this issue that there was unlikely to be any physical effects and the combination of the magnitude of the effects and probability of them occurring meant there was a negligible risk, which left a 'purely spiritual' element of risk that did not outweigh the scientific benefits of the work.[77]

Rightly or wrongly balancing such disparate factors can raise the risk that 'The ideology underpinning technological authoritarianism privileges 'scientific' rationality over social rationality or other values and epistemic frameworks'.[78] International legal obligations must also be observed. Getting the balance right will be difficult.

Where to Now?

Whether agricultural biotechnology can improve food security is hotly contested. Possible routes for assistance include increasing crop yield, expanding geographic zones suitable to grow food by improving the resilience of crops to environmental stresses such as heat, drought, flooding or salinity, and improving resistance to pests and disease. But there are highly contested issues about the role of biotechnology in food production when the focus widens to include SECs. Many European Union members and some developing countries in particular have been unwilling to adopt GM technology in agriculture or food because of such concerns. Opposition

76 New Zealand, Environmental Risks Management Authority Decision, Approval Code GMF000026 on Application Code GMF98009, 23 May 2001. See also discussion in Wheen, N. 2004. Genetic modification, risk assessment, and Maori belief under New Zealand's Hazardous Substances and New Organisms Act 1996. *Asia Pacific Journal of Environmental Law*, 8(3-4), 141-176.

77 Environmental Risks Management Authority (New Zealand), *Environmental Risks Management Authority Decision, Approval Code GMF000026 on Application Code GMF98009*, 23 May 2001, 21-22. [Online]. Available at: http://www.epa.govt.nz/search-databases/HSNO%20Application%20Register%20Documents/GMF98009%20Part%20II%20s67A%20-%202010.pdf [accessed: 15 February 2013].

78 Havemann, P. 2003. Genetic modification, ecological good governance and the law: New Zealand in the age of risk. *James Cook University Law Review*, 10, 7-50, 16.

to GM technology also comes from certain farmers, NGOs and members of the public and it can be expected that these differing views will be reflected in a nation's attitude to the role of GM science in addressing food security concerns. Regulatory frameworks relevant to agricultural biotechnology, whether in the context of the commercialization of GM crops, labelling for GM food or IP given to biotechnology developments, particularly where bioresources are sourced from indigenous or local communities, are already incorporating SECs. Whether this is appropriate depends partly on the costs and benefits of the inclusion, and the effect of that regulation on competition, its likely future effectiveness, compliance and administration burdens, and whether there are effective and efficient enforcement regimes, and an accountable body to administer it. Each country will respond differently to these matters.

However, the inclusion of SECs in agricultural biotechnology regulatory frameworks means that those frameworks are moving to reflect SECs of importance to the citizens of the nation concerned and how that nation considers the relevant SECs need to be balanced when responding to the adoption (or not) of biotechnology. This may be attractive because it enhances state sovereignty over food security for its people. But it has at least two important deficits.[79] First, the risk that inclusion of SECs, particularly if certain approaches to assessment are taken, may mean less democratic regulation of agricultural biotechnology because advocates can and have captured the process and other stakeholders with equally legitimate views are not given an equal opportunity to be involved. Second, there is the risk that the inclusion of SECs will make the regulatory system more complex and expensive and therefore more likely that only large multinational companies will be involved.[80] These outcomes are themself often cited as SECs justifying the rejection of agricultural biotechnology.

Further work also needs to be done on developing policy guidelines and methodology for including SECs in decision making. Without a clear and consistent technique, inclusion of SECs in national agricultural biotechnology regulatory frameworks will be vulnerable to claims that SECs are being used as unlawful restraints on trade contrary to international trade rules, even if that is not the case. As Falck-Zepeda et al. observe, SECs should not be and should not be seen to be being 'used (or abused) as a blanket justification for rejecting the technology without the need to support claims or debating the strengths of evidence presented to support such claims'.[81] There is also a risk that the introduction of SECs into

79 Strauss, D.M. and Strauss, M.C. 2009. Globalization and national sovereignty: Controlling the international food supply in the age of biotechnology. *Journal of Legal Studies in Business*, 15, 75-94, 76.

80 Tait, J. and Barker, G. 2011. Global food security and the governance of modern biotechnologies. *EMBO Reports*, 12(8), 763-768, 767.

81 Falck Zepeda, J., Wesseler, J. and Smyth, S. 2010. *The current status of the debate on socio-economic assessments and biosafety highlighting different positions and policies in Canada and the US, the EU and developing countries, Paper presented at 2010 World*

regulatory frameworks without justifiable and measured protocols will provide the opportunity for less scrupulous traders to claim certain SECs are addressed by their products or processes, when that may not be true (as is already happening with 'green washing'). A clear and consistent approach to SEC assessment and the proper definition of important terms helps avoid that exploitation occurring and builds trust between stakeholders. It also enables data and research to be more easily shared and developed.[82] This work is particularly important because in the context of agricultural biosafety, where there are as yet no international standards, guidelines or recommendations on assessment of SECs, any best practice developed under the CPB may become the de facto international protocol at least until and if it is successfully challenged in the WTO Dispute Settlement process.[83] The WTO Agreements have shown justifiable and measurably consistent approaches to regulation can be achieved. The move to include SECs in regulatory frameworks for agricultural biotechnology, though, highlights a trend to growing demands for more national sovereignty over issues as central as food security and perhaps a demand that the traditional recipe for international trade be changed. Done carefully, SEC inclusion may be an improvement to the recipe.

Congress of Environmental and Resource Economics, 8. [Online]. Available at: bch.cbd.int/cms/ui/forums/attachment.aspx?id=297 [accessed: 1 February 2013].

82 New Zealand ERMA. 1999. *Identifying Risks for Applications under the Hazardous Substances and New Organisms Act 1996*, ER-TG-01-1 9/99, Appendix B. [Online]. Available at: http://www.biosafety.gov.cn/image20010518/5068.pdf [accessed: 4 February 2013].

83 Smyth, S. and Falck-Zepeda, J. 2012. *Socio-economic Considerations and International Trade Agreements*, International Political Science Association Conference, Madrid, Spain, (8-12 July, 2012), 18. Copy on file with author.

PART IV
Intellectual Property, Food and Practice

Chapter 8

Can Intellectual Property Help Feed the World? Intellectual Property, the PLUMPYFIELD® Network and a Sociological Imagination

Jay Sanderson

Can intellectual property law help feed the world? Informed by C.W. Mills' 1959 sociological classic, *The Sociological Imagination*,[1] this chapter begins to answer the question of whether intellectual property law can help feed the world by directly relating and linking intellectual property law with the lives of those affected by food insecurity.[2] In exploring the role of intellectual property law in hunger, malnutrition and food insecurity, this chapter begins by suggesting that justifying intellectual property based on advances in technology is compromised by the complexity of food insecurity, and ultimately results in a *technology trap*; a situation in which intellectual property specifies food security as technology at the expense of the relations in which food is produced, distributed and accessed. The chapter then examines Nutriset's PlumpyField global supply network, an instance in which intellectual property is being used in the fight against hunger, malnutrition and food insecurity. In the context of food insecurity and intellectual property, then, the PlumpyField network provides a model in which intellectual property can be used as a vehicle for social change. More specifically, the PlumpyField network shows how intellectual property can be used to encourage and support local participation in the manufacture and distribution of food and nutritional products, and, thus, help create a local presence, help build local capacity, facilitate access to food and nutritional products, and ensure the quality of food and nutritional products (as well as broader health and safety standards) in developing countries.

---

1 Mills, C.W. 1959. *The Sociological Imagination*. New York: Oxford University Press.

2 Food insecurity is defined by the Food and Agriculture Organization as '[a] situation that exists when people lack secure access to sufficient amounts of safe and nutritious food for normal growth and development and an active and health life': Food and Agriculture Organization of the United Nations. 2010. *The State of Food Insecurity in the World 2010: Addressing Food Insecurity in Protracted Crises*. [Online]. Available at: http://www.fao.org/docrep/013/i1683e/i1683e.pdf [accessed: 4 February 2013].

The Technology Trap: How Intellectual Property Specifies Food Security as Technology

Proponents of robust intellectual property protection tend to rely on a direct connection between intellectual property and technological advancement to justify the grant of intellectual property in agricultural research and food production. More specifically, supporters of intellectual property law, particularly patents and plant variety rights, in agricultural research and food production argue that intellectual property is a catalyst for research investment and that they create a positive environment in which new, modified and improved technologies are developed. The traditional argument is: if researchers and companies can protect their innovations, and potentially recoup the costs of investment or even earn a profit, through the use of intellectual property rights, they are more likely to innovate and invent.[3] The connection between intellectual property and technological advancement is reflected in Article 7 of the *Agreement on Trade-related Aspects of Intellectual Property* (the 'TRIPs Agreement') which sets out the objectives of the intellectual protection and enforcement as 'contribut[ing] to the promotion of technological innovation and to the transfer and dissemination of technology, to the mutual advantage of producers and users of technological knowledge and in a manner conducive to social and economic welfare'.[4] The corollary of this argument is that, without intellectual property to incentivize agricultural research and food production, there would be less agricultural research, technological development and, ultimately, less food-related innovation.

But, in the context of food insecurity supporting intellectual property based on technological advancement in technology is compromised by the complexity of food insecurity, and by the fact that the link between intellectual property, technology and food security is based on a number of (often unstated) assumptions. One assumption that underpins the use of intellectual property in agricultural research and food production is that food insecurity is caused by a scarcity of food, as well as the related absence of high yielding, pest and climate resilient crops. If, however, we examine the rationale that technological advancement is necessary to overcome the problem of scarcity more closely, and with a sociological imagination, it becomes apparent that developing new technology and producing more food is not the panacea of food insecurity.[5] Lappé and Collins, who have been at the forefront of food policy critiques since the 1970s, argue that the source of world hunger is not a lack of food or technology but rather it

---

3 See, for example, Commission on Intellectual Property Rights. 2002. *Final Report of the Commission on Intellectual Property: Integrating Intellectual Property Rights and Development Policy*. [Online]. Available at: http://www.iprcommission.org/papers/pdfs/final_report/CIPRfullfinal.pdf [accessed: 4 February 2013].

4 *Agreement on Trade-related Aspects of Intellectual Property*, Article 7.

5 For an up-to-date and informative overview of the global agri-food system: see Carolan, M. 2012. *The Sociology of Food and Agriculture*. Oxon: Earthscan.

is colonialism and the subsequent exploitation of multinational companies that are the root causes of world hunger.[6] Also, in his book *Poverty and Famines: An Essay on Entitlement and Deprivation*, Amartya Sen examines a number of famines including the Bengal Famine of 1943, the Ethiopian Famine of 1972-4 and the Famine of Bangladesh in 1974, and argues that it is a loss of capability or entitlement rather than a scarcity of food that leads to famine. At the time, Sen's capabilities and entitlement approach challenged the Malthusian logic that there is simply too many people and not enough food, and shifted the focus of away from food production towards issues of ownership and exchange of food or other goods that provide the money to purchase food.[7] Significantly, Sen argues that in many cases of famine food supplies are not, in absolute terms, insufficient and on this point he is worth quoting at length:

> Sometimes famines have coincided with years of peak food availability, as in the Bangladesh famine of 1974. Even when a famine is, in fact, associated with a decline in food production, we still have to go beyond the output statistics to explain why it is that some parts of the population get wiped out, while the rest do just fine. Since food and other commodities are not distributed freely, people's consumption depends on their 'entitlements', that is, on the bundles of goods over which they can establish ownership through production and trade, using their own means. Some people own the food they themselves grow, while other buy them in the market on the basis of incomes earned through employment, trade or business. Famines are initiated by a severe loss of entitlements of one or more occupation groups, depriving them of the opportunity to command and consume food. Famines survive by divide and rule. For example, a group of peasants may suffer entitlement loss when food output in their territory declines, perhaps due to a local drought, even when there is no general dearth of food in the country. The victims would not have the means to buy food from elsewhere, since they wouldn't have anything much to sell to earn an income, given their own production loss. Others with more secure earnings in other occupations or in other locations may be able to get by well enough by purchasing food from elsewhere.[8]

---

6 Lappé, F. and Collins, J. with Fowler, C. 1977. *Food First: Beyond the Myth of Scarcity.* Boston: Houghton Miffin Company.

7 Sen, A. 1982. *Poverty and Famines: An Essay on Entitlement and Deprivation.* New York: Oxford University Press. More recently the complex mix of politics, water, land, energy and technology and knowledge in food insecurity was discussed in Cribb, J. 2010. *The Coming Famine: The Global Food Crisis and What we Can do to Avoid It.* California: University of California Press.

8 Sen, A. 1991. Public Action to Remedy Hunger. *Interdisciplinary Science Reviews*, 16(4), 324-336.

That food insecurity is a problem of access, as well as production, is exemplified by the so-called 'food crisis' of 2007-8 that occurred around the world in countries such as Morocco, Egypt, Italy, Haiti, Jordan, the Philippines, Argentina and Mexico.[9] In Mexico for example, tortillas (which have been a staple food in Mexico for thousands of years) became too expensive for much of the Mexican population. This was due to a range of complex, related factors. The *North American Free Trade Agreement* (NAFTA) entered into force in 1994 between the United States, Canada and Mexico to create the world's largest free trade area. While the objectives of the NAFTA include reducing barriers to trade in North America, improving working conditions in North America and creating a safe market for goods and services produced in North America, the NAFTA also tied Mexican corn production to the North American market. In the early 2000s, climate change became a driving force in international, regional and national policy and numerous countries introduced laws and regulation encouraging the use of agro-fuels. In 2005, the United States Congress, for example, established various measures to help reduce greenhouse gas emissions by encouraging the development and use of renewable fuels; leading, in 2007, to the introduction of the Renewable Fuels Standards program which set an agro-fuel mandate that encouraged the use of corn as fuel.[10]

Unsurprisingly, the demand for corn for use as a fuel increased substantially in the United States during the mid-2000s, leaving many farmers with little choice but to use their land for the production of corn for agro-fuel rather than for food.[11] As McMichael summarizes the situation: 'through rising demand for agro-fuels, and feed crops – to exacerbate food price inflation as their mutual competition for land has the perverse effect of rendering each crop more lucrative, as they also displace land used for food crops'.[12] In Mexico, the effect of this was that corn production moved away from contributing to staple foods such as tortillas and towards agro-fuels for the high value United States market. As a consequence Mexico, a country defined by corn, had to import corn from the United States

9 For a discussion of the 'food crisis' see Collier, P. The politics of hunger: how illusion and greed fan the food crisis. *Foreign Affairs*, 87(6), 67-79; Headey, D. and Fan, S. 2008. Anatomy of a crisis: the causes and consequences of surging food prices. *Agricultural Economics*, 39, 375-391.

10 See Tennenbaum, D. 2008. Food vs. fuel: Diversion of crops could cause more hunger. *Environmental Health Perspectives*, 116(6), A254-A257.

11 Anderson, J. and Coble, K. 2010. Impact of renewable fuels standard ethanol mandate on the corn market. *Agribusiness*, 26(1), 49-63; Cotula, L., Dyer, N. and Vermeulen, S. 2008. *Fuelling Exclusion? The Biofuels Boom and Poor People's Access to Land*. London: International Institute for Environment and Development.

12 McMichael, P. 2009. A food regime analysis of the 'World Food Crisis'. *Agriculture and Human Values*, 26(4), 281-295, 282.

and, as a consequence, the price of tortillas doubled during 2006.[13] So devastating has been the confluence of the NAFTA and the demand for agro-fuels on food supply that, on 9 August 2012 the Director General of the United Nations' Food and Agriculture Organization (FAO), José Graziano da Silva, called on the United States to suspend its ethanol mandate so that more crops, particularly corn, could be diverted back to food production.[14]

Another assumption that underpins and informs the use of intellectual property in agricultural research and food production is that intellectual property leads to crop improvement, and that ultimately this will lead to food security. Although the question of whether intellectual property law provides an incentive for technological development has been asked for decades, incentive-based rationales for intellectual property have proven difficult to substantiate.[15] Robert Merges sums up the problem of demonstrating that intellectual property provides an incentive to innovate or create in the following way:

> Estimating costs and benefits, modelling them over time, projecting what would happen under counterfactuals (such as how many novels or pop songs really would be written in the absence of copyright protection, and who would benefit from such a situation) – these are all overwhelmingly complicated tasks. And this problem poses a major problem for utilitarian theory. The sheer practical difficulty of measuring or approximately all the variables involved means that the utilitarian program will always be at best aspirational.[16]

In addition to the broader questions about whether intellectual property provides an incentive for technological development, the impact of technology on food insecurity is unclear. Although there have been attempts empirically to assess

13 Holt-Giménez, E. and Kenfield, I. 2009. When renewable isn't sustainable: Agrofuels and the inconvenient truths behind the 2007 U.S. Energy Independence and Security Act, in *Agrofuels in the Americas*, edited by R. Jonasse. Oakland: Food First Books, 25-40.

14 The World. 2012. *In Face of Drought, UN Asks US to Suspend Biofuel Mandate*, August 10, 2012. [Online]. Available at: www.theworld.org/2012/08/in-face-of-drought-un-asks-us-to-suspend-biofuel-mandate [accessed: 4 February 2013].

15 In relation to patents see, for example, Lerner, J. 2009. The empirical impact of intellectual property rights on innovation: Puzzles and clues. *The American Economic Review*, 99(2), 343-348; Varsakelis, N. 2001. The impact of patent protection, economy openness and national culture on R&D investment: A cross-country empirical investigation. *Research Policy*, 30(7), 1059-1068. In relation to plant variety rights see, for example, Tripp, R., Louwaars, N. and Eaton, D. 2007. Plant variety protection in developing countries: A report from the field. *Food Policy*, 32(3), 354-371; International Union for the Protection of New Varieties of Plants. 2005. *Report on the Impact of Plant Variety Protection*. Geneva: UPOV.

16 Merges, R. 2011. *Justifying Intellectual Property*. Cambridge, MA: Harvard University Press, 2-3.

the impact of technology on food security, this, in a similar way to proving that intellectual property law provides an incentive for technological advancement, has proven difficult.[17] In keeping with a sociological imagination when thinking about the effect of technology on food security, it is helpful to consider the people affected by technological advances in agriculture and food production rather than merely focusing on the technological advances related to agriculture and food production. In this way the so-called Green Revolution is an ideal site to scrutinize the effect of technological advances and crop improvement on food security.

First used in 1968 by William Gaud, the Director of the United States Agency for International Development, the phrase 'the Green Revolution' refers to a period of technological advancement that began in the 1940s and lasted until the late 1970s. This period was characterized by the modernization of agriculture through a suite of measures including improved crop varieties, irrigation, fertilizers, pesticides, extension services and the availability of credit to famers.[18] There is no denying that the Green Revolution had a significant impact on agriculture, for example: wheat production in Pakistan increased by 60 per cent from 1967 to 1969; India achieved self-sufficiency in cereals in the mid-1970s; and Indonesia's rice yields nearly doubled from 1965 to 1989.[19] As one commentator noted: 'history records no increase in food production that was remotely comparable in scale, speed and duration'.[20] Despite the increases in food production, the impact of the Green Revolution on food security is less clear. Critics of the Green Revolution question whether technological advances – such as improved corn, wheat and rice varieties – resulted in a significant and lasting impact on food security. A review of 300 studies on the Green Revolution published between 1970 and 1989 shows that the effect of the Green Revolution's improved crop varieties was inconsistent at best: in fact, 80 per cent of the articles reviewed concluded that the Green Revolution increased inequality.[21] More specific condemnation of the Green Revolution has focused on the fact that to be effective the technology developed required considerable resources such as fertilizers, pesticides, irrigation and machinery; all of which

17 Kasturi, P. 2009. Technology and food security. *Humanomics*, 25(2), 163-168; Singh, P. 2005. Technology options for achieving food security in South Asia, in *Economic Reforms and Food Secuirty: The Impact of Trade and Technology in South Asia*, edited by S. Babu and A. Gulati, New York: Food Products Press, 163-173; Oehmke, J. and Crawford, W. 1996. The impact of agricultural technology in sub-Saharan Africa. *Journal of African Economies*, 5(2), 271-292.

18 For an overview of the Green Revolution see Conway, G. 1997. *The Doubly Green Revolution*. London: Penguin.

19 Pinstrup-Andersen, P. and Hazell, P. 1985. The impact of the green revolution and prospects for the future. *Food Reviews International*, 1(1), 1-25.

20 Lipton, M. 1989. *New Seeds and Poor People*. Baltimore: Johns Hopkins University Press, 1.

21 Freebairn, D. 1995. Did the Green Revolution concentrate incomes? A qualitative study of research papers. *World Development*, 23(2), 265-279.

are not always readily available in developing countries.[22] Furthermore, uptake of technology is dependent on policy, the availability of loans, credit or subsidies, and the availability of the necessary infrastructure such as roads, railways and ports.[23] According to critics, the Green Revolution also assumes a one-size-fits-all approach, and, as a consequence, had little impact on developing countries such as Africa which had institutional and structural problems, political and economic instability, corruption and an absence of resolve to make meaningful, long-term change.[24]

When thinking about the role that intellectual property law plays (or might play) in feeding the world it is important to keep in mind that food insecurity is a complex physical, economic, political and social problem. Too often, however, it is assumed that crop improvement is the panacea of food insecurity, and, thus, intellectual property is justified on the grounds that it provides the necessary incentives for crop improvement and food production. While this might be true in some circumstances – and there are a range of technological innovations such as transgenic crops, fertilizers and pesticides that can potentially increase agricultural productivity – merely focusing on the relationship between intellectual property and technological advancement leads to a *technology trap*: a situation in which food security is specified as technology at the expense of the relations in which food is produced, accessed and distributed. Avoiding the *technology trap* is particularly important because, despite advances in technology and food production, hundreds of millions of people are hungry not because of an absolute scarcity of food but because they lack the means to produce or purchase the food that they need for a healthy and productive life.[25] With this is mind, the remainder of this chapter examines Nutriset's PlumpyField global supply network of food and nutritional products, an example of intellectual property being used to encourage and support the participation of entities in developing countries in the manufacture of food and nutritional products.

---

22 Shiva, V. 1992. *The Violence of the Green Revolution: Third World Agricutlure, Ecology and Politics*. London: Zed Books; Mechlem, K. 2010. Agricultural biotechnologies, transgenic crops and the poor: Opportunities and challenges. *Human Rights Law Review*, 10(4), 749-764.

23 See, for example, Thurow, R. and Kilman, S. 2010. Fighting famine: A band-aid for the poor has become an industry for the rich. *Human Rights* 37(1), 10; Mandell, B. 2009. Cultivating race: How the science and technology of agriculture preserves race in the global economy. *Albany Law Review*, 72(4), 939-951.

24 See, for example, Evenson, R. and Gollin, D. 2003. Assessing the impact of the Green Revolution, 1960 to 2000. *Science*, 300(5620), 758-762.

25 Food and Agriculture Organization of the United Nations. 2012. *Towards the Future We Want: End Hunger and Make the Transition to Sustainable Agricultural and Food Systems*. Rome: FAO.

The PLUMPYFIELD® Global Supply Network: Using Intellectual Property to Encourage and Support Local Participation

Established in 1986, Nutriset is a for-profit company based in France that specializes in developing food and nutritional products that are used to prevent and treat malnutrition and diarrhoea in developing countries. Nutriset is best known for developing, in partnership with the Institut de Recherché pour le Développement (IRD), the first peanut-based ready-to-use therapeutic food (RUTF) for the treatment of severe acute malnutrition known as Plumpy'Nut. Plumpy'Nut comes in a 92 gram foil packet, provides around 500 kcal and is enriched with vital vitamins and minerals. It does not need refrigeration, cooking or water, and lasts up to two years. So effective is Plumpy'Nut in treating acute malnutrition that it is one of the most widely used therapeutic solutions to fight child malnutrition and is used by a number of world health agencies including the United Nations Children's Fund (UNICEF) and the World Food Programme (WFP). In addition to Plumpy'Nut, Nutriset have developed a wide range of food and nutritional products including Plumpy'Doz, Plumpy'Sup, Nutributter, QBmix and Plumpy'Soy.[26]

In response to an increase in demand for Plumpy'Nut, as well as pressure to permit local production of Plumpy'Nut, Nutriset's food and nutritional products and services are distributed to developing countries largely through the PlumpyField network.[27] Formalized in 2005, the PlumpyField supply network is a franchising scheme that provides developing countries with access to Nutriset's products as well as information and know-how about technical matters such as production processes, quality control and management, and staff training. Importantly, the PlumpyField network, and the more recent ZincField network, have been established to allow Nutriset to monitor and control the production of its food and nutritional products.[28] By having a high level of involvement and control over the manufacture of their products, Nutriset contend that they are able to do more than merely supply food and nutritional products to those in need. More specifically Nutriset argue that through the PlumpyField network they are able to exclude companies in developed countries from producing Nutriset's products, and can instead, encourage and support local partners in the (developing) countries to produce their own food and nutritional products. In so doing, Nutriset argue that they are increasing the economic and social effect of their products in developing

26 Nutriset. 2012. *Nutriset's Product Range*. [Online]. Available at: http://www.nutriset.fr/en/product-range/nutriset-product-range.html [accessed: 4 February 2013].

27 Guimón, J. and Guimón, P. 2012. How ready-to-use therapeutic food shapes a new technological regime to treat child nutrition. *Technological Forecasting & Social Change* 79(7), 1319-1327.

28 Launched in 2010, the ZincField global supply network has been established by Nutriset to help treat acute diarrhoea; as of 21 November 2012, the ZincField network has two partners, one in Ethiopia and the other in Tanzania.

countries. In 2012, 12 of the 13 PlumpyField manufacturing partners were based in developing countries (see Table 8.1).

Table 8.1 PlumpyField Members

Manufacturing Partner	Country	Year in which the Manufacturing Partner Joined the PlumpyField Network
Société de Transformation Alimentaire (STA)	Niger	2005
Hilina Enriched Foods	Ethiopia	2006
Joint Aid Management (JAM)	Mozambique	2007
Vitaset	Dominican Republic	2008
Tankjaka Food	Madagascar	2009
Power Foods	Tanzania	2009
Edesia	United States of America	2009
Meds & Foods for Kids	Haiti	2010
NutriVita Food	India	2010
InnoFaso	Burkino Faso	2011
Reco Industries	Uganda	2011
Samil Industrial Co	Sudan	2011

Note: Nutriset. 2012. *Members of the PlumpyField Network*. [Online]. Available at: http://www.nutriset.fr/en/international-networks/plumpyfield-network/members-of-the-plumpyfield-network.html [accessed: 21 November 2012].

In addition to the PlumpyField and ZincField networks, Nutristet uses online *Patents Usage Agreements* to grant non-exclusive licences to entities in developing countries.[29] The aim of the *Patents Usage Agreements* is to 'allow

29 Nutriset. 2012. *Patents Usage Agreement*. A copy of the *Patents Usage Agreement* is included in the Appendix (permission to include the Patents Usage Agreement was

companies in developing nations to benefit from [Nutriset's] common patents and, for this purpose, enter into Usage Agreement via a simplified internet-based system'. To be eligible for the *Patents Usage Agreement* entities must be local non-government organizations, and must 'have its production and business site, activity, headquarters, and main shareholders (at least 51% of the share capital) based' in one of the developing countries in which Nutriset has valid patent rights.[30] According to Nutriset, these countries include Benin, Burkina Faso, Cameroon, Cote d'Ivoire, Gabon, Gambia, Ghana, Guinea, Guinea-Bissau, Equatorial Guinea, Kenya, Lesotho, Malawi, Mali, Mauritania, Uganda, Central African Republic, Republic of the Congo, Senegal, Sierra Leone, Sudan, Swaziland, Tanzania, Chad, Togo and Zimbabwe. While the 'beneficiaries' of the *Patents Usage Agreement* are granted non-exclusive authorization to use Nutriset's patents, they do not receive the associated benefits of joining the Plumpy'Field network. More specifically, in return for being permitted to use the processes and methods needed to prepare (similar) products to those developed by Nutriset the licensee must 'develop its own formulae, recipes and Products, without assistance from Nutriset and the IRD',[31] 'undertake to develop and implement its own quality system'[32] and not 'reproduce and/or imitate the brands, logos, packaging, graphic identity or any other distinctive signs used by Nutriset for its own products, packaging and communication media'.[33]Whilst it is not compulsory for the patent licensee to pay a licence fee they are 'invited' to make a 1 per cent contribution of the turnover earned by the sales of the products included in the *Usage Agreement* to fund the IRD's research and development.[34] As of 10 December 2012 six entities have signed up to the *Patents Usage Agreement* (see Table 8.2).

One of the reasons why Nutriset is able to adopt a franchise and licensing system is because they hold intellectual property over some of their products, names and brands.[35] The most well-known of Nutriset's intellectual property is

granted by Priscilla Hoarau, Legal & Valorisation Department, Nutriset (11 December 2012)).

30 Nutriset. 2012. *Nutriset/IRD's Patents Usage Agreement.* [Online]. Available at: http://www.nutriset.fr/en/access/patents-for-development/online-patent-usage-agreement.html [accessed: 21 November 2012].

31 Nutriset. 2012. *Patents Usage Agreement*, Article 2.3.

32 Nutriset. 2012. *Patents Usages Agreement*, Article 2.4.

33 Nutriset. 2012. *Patents Usages Agreement*, Article 5.1.

34 The co-creators of the Plumpy'nut are listed as André Briend (a scientist at the Institut de Recherché pour le Développement (IRD)) and Michel Lescanne (the Chief Executive Officer of Nutriset), although Nutriset is the sole assignee, and administers the patents.

35 In 2010, Nutriset were also involved in a successful domain name dispute over the use of the term 'plumpynut' in the domain name <plumpynutinthefild.com>: WIPO

Table 8.2 Parties to Nutriset's Patents Usage Agreement

Parties to Nutriset's Patents Usage Agreement	Country	Year in which Party Agreed to the Patents Usage Agreement
Les Moulins du Sahel	Mali	2010
Project Peanut Butte	Sierra Leone	2011
Kamili Packers LTD	Kenya	2011
ETS Agrisol	Chad	2011
INSTA Products (EPZ) LTD	Kenya	2011
Project Peanut Butter	Malawi	2012
Nkhoma Hospital	Malawi	2012
Valdafrique Laboratories Canonne	Senegal	2012

Note: Nutriset. 2012. *Beneficiaries of a Patents Use Agreement*. [Online]. Available at: http://www.nutriset.fr/en/access/patents-for-development/online-patent-usage-agreement/beneficiaries-usage-agreement.html [accessed: 21 November 2012].

the Plumpy'Nut patent which is registered with the World Intellectual Property Organization (WIPO) with the title *High Energy Complete Food or Nutritional Supplement, Method for Preparing Same and Uses Thereof.*[36] The Plumpy'Nut patent has been registered with the French patent office, the Canadian Intellectual Property Office, the European Patent Office and the United States Patent and Trademark Office. Perhaps most importantly, however, Nutriset has strategically sought, and been granted, patents in a number of developing countries including Benin, Burkina Faso, Cameroon, Cote d'Ivoire, Gabon, Gambia, Ghana, Guinea, Guinea-Bissau, Equatorial Guinea, Kenya, Lesotho, Malawi, Mali, Mauritania, Uganda, Central African Republic, Republic of the Congo, Senegal, Sierra Leone, Sudan, Swaziland, Tanzania, Chad, Togo and Zimbabwe.[37] Somewhat contentiously, the Plumpy'nut patent is broadly written and covers a number of

Arbitration and Mediation Center, *Nutriset v. Private Registrations Aktien Gesellschaft*, Case No. D2010-0358 (29 April 2010).

36 Nutriset, *High Energy Complete Food or Nutritional Supplement, Method for Preparing Same and Uses Thereof*, World Intellectual Property Organization (Publication No. WO2002034077).

37 These countries are listed by Nutriset on their website, and are included in the *Patents Usage Agreement*. See, Nutriset. 2012. *Nutriset/IRD's patents usage agreement.*

products as well as processes for making Plump'nut. The Plumy'nut patent has 23 claims that include, for example, a:

> [c]omplete food or nutritional supplement which contains at most 10% by weight of water, develops an osmolality of less than 100 mOsm/kg after immersion in four times its own volume of water and is stable to oxidation, comprising a mixture of food-grade products, said mixture being coated with at least one lipid-rich substance derived from oleaginous seeds and being enriched in vitamins, soluble or insoluble mineral salts, enzymes or mixtures thereof; [and a]
>
> [p]rocess for the preparation of a complete food or food complement according to claim 1, wherein all or part of the products which provide carbohydrates, the products which provide proteins, and the vitamins, minerals, enzymes and additives is mixed, the products which provide lipids and the coating substance are then added to this mixture, the ingredients are mixed again and the resulting product is then poured into the pack.[38]

In addition to holding patent rights in both developed and developing countries Nutriset have several registered trade marks. The names PlumpyField, ZincField, Plumpy'Nut, Plumpy'Doz and Nutributter are registered trade marks in a range of jurisdictions world-wide, including the Unites States and some developing countries. The word PlumpyField, for example, is registered under International Classification 35 (for goods and services related to business management and consultancy, particularly for a franchisor in the field of food production and preparation intended for the prevention and the fight against malnutrition) and the word Plumpy'Nut is registered under International Classification 5 (for food and substances adapted for medical uses, and dietary supplements for human consumption).

One of the defining features of intellectual property is that they enable the owner to control who has access to the products, processes or brands that are protected. Indeed, Nutriset's franchise and licensing strategy is largely made possible by the fact that they have sought, and been granted, intellectual property in both developed and developing countries. By proactively using and defending its intellectual property, Nutriset is able to exert control over whom, and under what circumstances, their products and services are used. Generally speaking, manufacturers only obtain permission to use Nutriset's registered patents and trade marks once they have joined the PlumpyField network. Take, for example, the Indian-based company NutriVita Food which was formed in 2010 to 'eradicate under-nourishment and create a healthy and well nourished younger generation'

[Online]. Available at: www.nutriset.fr/en/access/patents-for-development/online-patent-usage-agreement.html [accessed: 4 February 2013].

38 Nutriset, *Food or Nutritional Supplement, Preparation Method and Uses*. United States Patent and Trademark Office (Patent No. 6,346,284). Issued on 12 February 2002.

in India. NutriVita Food became a PlumpyField member in 2010 and in so doing obtained permission from Nutriset to manufacture and distribute Nutriset's patented products – including Plumpy'Nut, Plumpy'Sup and Plumpy'Doz. As a PlumpyField member, NutriVita Food can also use the associated PlumpyField, Plumpy'Nut, Plumpy'Sup and Plumpy'Doz trade marks, packaging and branding (see Figures 8.1 and 8.2).

Figure 8.1 Plumpy'Nut

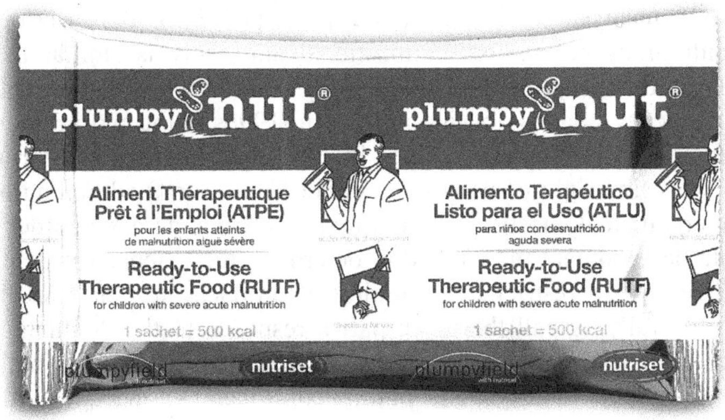

Figure 8.2 Plumpy'Doz

Building Nutritional Autonomy by Creating a Local Presence, Increasing Local Capacity, Facilitating Access and Ensuring the Quality Food Products

By using their patent and trade mark rights to control access to their products, processes and brands Nutriset hopes to encourage and support local participation in the production and distribution of its food and nutritional products. Indeed, one of the underlying principles of the PlumpyField global network, as well as Nutriset's *Patents Usage Agreement*, is nutritional autonomy. Defined by Nutriset as '[t]he capacity of a country or community to set up a sustainable system to identify and make accessible the nutrients required for the development and good health of its population',[39] nutritional autonomy is similar to the more commonly used concept of food sovereignty. First proposed by the international farmers' advocacy, Via Campesina in the mid-1990s at the World Food Summit, food sovereignty – which was defined in 2007 at an international forum on food sovereignty as 'the right of peoples to define their own food and agriculture; to protect and regulate domestic agricultural production and trade in order to achieve sustainable development objectives; to determine the extent to which they want to be self-reliant; to restrict the dumping of products in their markets'[40] – has come to be a rallying call that is in part a response to the dominance of large corporate-driven food production and the proliferation of free trade policies. With its emphasis on local food systems, putting control and knowledge in the hands of local farmers and local manufacturers, food sovereignty movements are seen as long-term solutions to alleviate hunger, malnutrition and food insecurity.[41]

The significance of locally oriented solutions to the long-term alleviation of hunger, malnutrition and food insecurity is underlined by criticisms of alternative forms of assistance that are provided to developing countries. One form of assistance, food aid, is a multifaceted approach that is largely driven by international donors and international organizations such as the United Nations' WFP and FAO. Food aid is often provided in the form of in-kind or subsidized deliveries of food and nutritional products. Traditionally, one of the main strategies of international food aid has been to export food and nutritional products from developed countries

---

39 Nutriset. 2012. *Nutritional Autonomy for All*. [Online]. Available at: http://www.nutriset.fr/en/access/nutritional-autonomy.html [accessed: 4 February 2013].

40 Forum on Food Sovereignty. 2007. *Peoples' Food Sovereignty Statement*, 1. [Online]. Available at: http://www.nyeleni.org/spip.php?article147 [accessed: 4 February 2013]. The Forum on Food Sovereignty brought together over 700 NGOs, CSOs, environmentalists, indigenous people, farmers, fisherfolk and trade unions.

41 For a discussion of food sovereignty, see Desmarais, A., Wiebe, N. and Wittman, H. 2010. *Food Sovereignty: Reconnecting Food, Nature and Community*. Black Point: Fernwood Publishing Company Limited; Boyer, J. 2010. Food security, food sovereignty and local challenges for transnational agrarian movements: The Hounduras case. *The Journal of Peasant Studies*, 37(2), 319-351; Edwards, F. 2010. Food sovereignty and social change in Venezuela. *Chain Reaction*, 109, 21-23.

to those countries in need.[42] Another form of assistance increasingly being used to respond to hunger, malnutrition and poverty is the social business model known as 'buy one give one'. The 'buy one give one' model is an approach in which companies donate products or services to countries or individuals in need every time a customer buys one of their products. One of the largest and most well-known 'buy one give one' businesses is Toms shoes; a United States based company founded in 2006 that, for every pair of shoes that someone buys, gives a pair shoes to a child in need.[43] An example of a food-based 'buy one give one' business is Two Degrees of Food. Based in the United States, Two Degrees of Food has the slogan 'Buy a Bar: Give a Meal', and, have, according to their own website, 'committed' over 500,000 meals to those in need.[44]

While food aid and donation programs such as 'buy one give one' assist with easing the immediate problem of hunger and the treatment of acute malnutrition, they are less likely to provide long-term, sustainable solutions to food insecurity. These approaches have a tendency to focus on the short-term supply of life-saving energy and nutrients, but are seldom customized to the underlying causes of the problems facing the people being aided. This means that food aid and donation programs can impose ill-fitting solutions on communities. Furthermore, food aid and donation programs are often tied to conditions stipulated by the donor that, rather than support food sovereignty or nutritional autonomy, promote the domestic interests of donor countries, organizations or businesses.[45] The United States, the biggest donor of food aid, for example, commonly ties its food aid to the interests of United States farmers, United States agribusiness and the United States transport industry.[46] Generally speaking, this means that the food and nutritional products donated by the United States are purchased in the United States, packaged in the United States and shipped to the country in need – often to the advantage of United States businesses.[47] By favouring the interests of the

---

42 Barrett, C. and Maxwell, D. 2005. *Food Aid After Fifty Years: Recasting its Role*. London: Routledge; Shaw, J. 2001. *The UN World Food Programme and the Development of Food Aid*. New York: Palgrave.

43 Other 'well-known' buy one give one businesses are Warby Parker (who give a pair of prescription glasses every time a customer buys a pair) and the Human Time Project (in which for every watch sold another is given to a health care worker that 'needs one').

44 Two Degrees. 2012. *Our Mission*. [Online]. Available at: http://twodegreesfood.com/mission [accessed: 4 February 2013].

45 Mousseau, F. 2005. *Food Aid or Food Sovereignty? Ending World Hunger in Our Time*. [Online]. Available at: http://www.oaklandinstitute.org/sites/oaklandinstitute.org/files/fasr.pdf [accessed: 4 February 2013].

46 Indeed, in the United States, food aid is legislatively tied to United States Agriculture: see, for example, Cooper, A. 2011. Peanuts, Politics and the Plumpy'nut Patent. *European Intellectual Property Review*, 33(8). 481-489.

47 Riddell, R. 2007. *Does Food Aid Really Work?* Oxford; New York: Oxford University Press.

donors, food aid and donation programs can undermine local producers, and can cause local markets to be uncompetitive. Ultimately, then, there is minimal long-term improvement in food security by giving away free or cheap imported food and nutritional products. Significantly, some of the aid and donation programs have acknowledged that there are problems associated with free, cheap and/or subsidized food, with many food aid programs shifting their attention to growing the participation of local communities and peoples. In so doing some of the aid and donation programs are purchasing local food and nutritional products from local communities, while others are becoming more focused on local capacity building.[48]

In light of the criticisms of food aid and donation programs, the advantages of using intellectual property to support locally oriented solutions becomes evident. One benefit of using intellectual property to support local participation is that it creates a presence in the local community. By creating a local presence, farmers, producers and consumers are brought together in a way that benefits the whole community, and in a way that is more likely to lead to the long-term alleviation of hunger, malnutrition and food insecurity. Under the Plumpy'Field network: local farmers benefit by selling produce such as peanuts, oil and sugar to the local manufacturers; local manufacturers benefit from gaining access to Nutriset's knowledge, skills and practices, and, as a result, are able to produce products that are of a consistently high quality; local people benefit from increased employment, fair wages and good working conditions; local consumers benefit from having access to locally produced food and nutritional products. Nutriset estimate that the PlumpyField network has contributed to the employment of over 500 individuals in developing countries such as Haiti, Ethiopia and India.[49] Nutriset also estimate that PlumpyField members have purchased over 800 tons of peanuts, over 870 tons of edible oils and over 970 tons of sugar, all from local farmers.[50]

Another advantage of creating a local presence is that it helps to build and strengthen local capacity, or to use Sen's terminology, 'entitlements'. Building local capacity is perhaps the most important aspect of Nutriset's use of intellectual property rights: according to the FAO, improving local production by small-scale farmers is one of the best ways of ensuring food security because it provides income and employment for local people.[51] Under Nutriset's franchise model, the PlumpyField members in developing countries act as franchisees to produce

48 Barrett, C. and Maxwell, D. 2005. *Food Aid After Fifty Years: Recasting its Role*. London: Routledge.

49 Nutriset. 2012. *Key Figures* [Online]. Available at: http://www.nutriset.fr/en/about-nutriset/2011-key-figures.html [accessed: 4 February 2013].

50 Ibid.

51 Food and Agriculture Organization of the United Nations. 2010. *The State of Food Insecurity in the World: Addressing Food Insecurity in Protracted Crises*. [Online]. Available at: http://www.fao.org/docrep/013/i1683e/i1683e.pdf [accessed: 4 February 2013].

and distribute Nutriset products and they also 'benefit from the enhancements and innovation to these same types of products'.[52] PlumpyField members also benefit from having access to Nutiset's information and know-how, and by being trained in technical matters such as production processes, quality control and pest management.

Together, the local presence and capacity building strengthen local markets and discourage non-local companies from importing similar products into developing countries. The experience with food aid suggests that the existence of handouts, or cheap imported food and nutritional products, may distort local markers and prevent local industry from flourishing. This means that if Nutriset's products are manufactured without restriction in developing countries and are then shipped to other developing countries, then to the extent there are local producers and manufacturers, they face greater (often subsidized) competition before they can have successful businesses that contribute to the local economy and create local employment. Furthermore, if companies in the developed world are able to drive down the price of products such as Plumpy'Nut, this may create disincentives for longer term investments, discourage domestic food production, hurt local peanut growers, workers and manufacturers, and ultimately cause long-term harm to developing countries.[53] Referring to Nutriset's much sought after Plumpy'Nut, Adeline Lescanne, Deputy General Manager of Nutriset, states:

> Some may pretend they are able to produce the equivalent of Plumpy'Nut at a cheaper price, but we fear that those solutions may not be [long lasting]. What should be the goal: to have companies manufacturing an RUTF in the North or to have them helping to develop local nutrition capacities, working with local health authorities, transferring competences to the South?[54]

Yet another benefit of Nutriset's locally-oriented approach is that it facilitates access to important and necessary food and nutritional products. The idea that Nutriset's use of intellectual property may improve access to life-saving food and nutritional products is inconsistent with the sentiments of opponents to Nutriset's approach. Critics of the Plumpy'Nut patent argue that intellectual property restricts access to life-saving products because they grant a monopoly over vital products and processes that reduces competition for those products and processes. In 2009,

52 Nutriset. 2012. *From 'Feed the Children' to Nutritional Autonomy for All,* 29. [Online]. Available at: http://www.nutriset.fr/assets/Rapports-annuels/rapport-nutriset-2011-web.pdf [accessed: 4 February 2013].

53 See, for example, FAO Geneva Round Table on Selected Issues, in particular Food Aid, 30 November 2001. [Online]. Available at: http://www.fao.org/economic/est/est-events-new/trade-meetings/fao-geneva-round-tables-symposia-and-workshops/en/ [accessed: 4 February 2013].

54 Cited in Dyer, C. 2010. Two charities challenge company's patent on Plumpy'Nut. *British Medical Journal*, 340(7755), 1055.

two United States based not-for-profit companies, Breedlove Foods and Mama Cares Foundation, sought a declaration that, by producing their own peanut-based RUTF called Re:Vive, they were not infringing Nutriset's United States patent *Food or Nutritional Supplement, Preparation Method and Uses* (Patent Number 6,346,284). The plaintiffs also claimed that Nutriset's patent is invalid because it is not novel or non-obvious.[55] To be successful in such a claim the Breedlove Foods and Mama Cares Foundation had to satisfy the court that there was a 'substantial controversy'.[56] In this case the plaintiffs claimed that Nutriset had declined to licence the Plumpy'Nut patent to third parties and that they had also 'warned' other companies that they would not tolerate infringement of the Plumpy'Nut patent.[57] In finding for Nutriset, the United States District Court concluded that Breedlove Foods and Mama Cares did not show a 'substantial controversy' required by the relevant law, because the 'plaintiff's complaint fails to allege any contact between plaintiffs and defendant prior to litigation, informal or otherwise' and because 'these facts are insufficient to establish an actual controversy'.[58]

While the dispute between Mama Cares, Breedlove Foods and Nutriset was decided on a technical legal issue, and, therefore, the United States District Court did not comment on the appropriateness of Nutriset's use of its patent rights, it does raise the question about whether intellectual property hinders access to necessary food and nutritional products. On the one hand, media and commentary on the dispute portray it as one between two charities and a money-driven corporation, where intellectual property is being used by the corporation to hinder the production of the life-saving Plumpy'Nut.[59] In a similar way, Breedlove and Mama Cares were quick to claim that Nutriset was limiting access to life-saving peanut-based nutritional supplements, and that Nutriset is more concerned with avoiding competition and enhancing its profits than humanitarian needs. On the other hand, it can be argued that using intellectual property to support the local production and manufacture of food and nutritional products facilitates, rather than hinders, access to food and nutritional products, and in the long term this will have a bigger impact on hunger, malnutrition and food insecurity. While the bulk

55 Cooper, above n 46.

56 The claim was made under the *Declaratory Judgment Act* 28 U.S.C., § 2201(a).

57 Medecins Sans Frontiers. 2009. *MSF Letter to Nutriset on Intellectual Property*. [Online]. Available at: http://www.msfaccess.org/sites/default/files/MSF_assets/MalNut/Docs/NUT_letter_NutrisetPatent_ENG_2009.pdf [accessed: 4 February 2013]. The open letter was sent by Dr Tido von Schoen-Angerer, Director of Medecins Sans Frontiers' Access to Essential Medicine Campaign.

58 *Mama Cares Foundation v Nutriset Société Par Actions Cimplifiée*, United States District Court for the District of Columbia (18 November 2012).

59 See, for example, Arie, S. 2010. Hungry for profit. *British Medical Journal*, 341(7776), 758-759; Schofield, H. 2010. *Legal Fight over Plumpy'Nut, The Hunger Wonder Product*. [Online]. Available at: www.news.bbc.co.uk/2/hi/europe/8610427.stm [accessed: 4 February 2013].

of Plumpy'Nut is still being produced in France, in 2011, according to Nutriset, 11,676 metric tons (out of a total of 46,000 metric tons) of Plumpy'Nut were made in developing countries such as Haiti (by Meds & Food for Kids), Ethiopia (by Hilina Enriched Foods), India (by NutriVita Foods) and Sudan (by Samil Industrial Company).[60]

A locally oriented approach can help overcome the difficulty of reaching the target population in nations with poor infrastructures, political instability and conflict that characterize many food insecure countries. Locally sourced and produced products are more accessible for a number of reasons. First, the food and nutritional products produced by PlumpyField members do not have far to travel in order to reach its destination. This means that there are fewer delays between the production and distribution of the products, as well as lower transportation and administration costs. It also means that the local manufacturer can be an integral part of local public health programs. In addition to the logistical advantages of locally-oriented production and distribution of food and nutritional products there are also broader political and social reasons why access to food and nutritional products might be improved through local production. Predictably, in countries suffering from famine and malnutrition, food is a valuable commodity that is often fought over. This means that access to imported food aid and food supplies is sometimes disrupted by militia groups and political leaders in order to exert control, gain a financial advantage or even to show their abhorrence of developed countries such as the United States. Alarmingly, a number of humanitarian aid workers, from organizations such as Doctors Without Borders and the World Food Programme, have been attacked and killed by militia groups, armed soldiers or individuals in countries suffering the worst famines such as Sudan, Ethiopia, Iraq and Somalia.[61] In Somalia, in 2012, the International Committee of the Red Cross was forced to suspend its food aid due to the dangers presented by the Islamish Shebab insurgence. Militia groups and political leaders have also affected shipments of Nutriset's Plumpy'Nut: in July 2011, for example, an aid flight carrying enough Plumpy'Nut to feed 3,500 children for a month arrived in Somalia but its distribution was unreliable because of the influence of the Islamish Shebab and their belief that food and nutritional aid is supplied largely by the United States, a country viewed by the Somalia Militia as anti-Muslim.[62]

Yet another benefit of Nutriset's locally oriented approach is that Nutriset is able to ensure that the processes used and products produced are of a high standard. Indeed, Nutriset supplies PlumpyField members with the essential minerals and vitamins used in the form of a premix so that local manufacturers can produce

60 *Key Figures*, above n 49.

61 See, for example, The Editor. 2011. The Horn of Africa: Chronicle of a famine foretold, did the world react too late to signs of famine in Somalia? *The Economist*. [Online, 30 July]. Available at: http://www.economist.com/node/21524864 [accessed: 4 February 2013].

62 See ibid.

food and nutritional products that are of a consistent, reliable and effective quality. By providing the necessary vitamins and minerals, as well as information, know-how and guidance on quality assurance, food handling, food safety and pest management, Nutriset has a direct, positive impact on local manufacturers. While a patent right does not guarantee the quality of the patented product or process, there is a body of research showing that Plumpy'Nut is effective in treating severe acute malnutrition; with a success rate of over 80 per cent.[63] As a consequence of the direct involvement of Nutriset in the manufacturing process, many of the local manufacturers sell their locally sourced and produced products to international, regional and national aid agencies, as well as to local clinics and suppliers. Furthermore, the associated trade marks act as a 'badge of origin' so that aid agencies, and ultimately end-users, are assured that they are using 'real' Nutriset products that meet their expectations and that have been scientifically shown to be effective. In 2012, for example, Haitian-based Meds & Food for Kids entered into a long-term agreement with the UNICEF so that its version of Plumpy'nut, Medika Mamba, will be purchased by UNICEF and used to treat child malnutrition in Haiti. In order to reach this agreement with UNICEF, Meds & Foods for Kids had to meet various quality standards and targets, Meds & Food for Kids acknowledge that becoming a PlumpyField member made this process easier.

In addition to directly ensuring the quality of the products produced by local manufacturers, Nutriset's involvement with local manufacturers has a secondary benefit for local manufacturers as they are better placed to be involved in other projects. This is achieved in part because the manufacturing partners, and their employees, develop a broader set of knowledge and skills around production processes, quality control, and health and safety procedures. In 2011, for example, Meds & Food for Kids were awarded a competitive United States Department of Agriculture grant to develop and implement a school snack program through Haiti's National School Lunch Program. Meds & Food for Kids believe that this program will decrease the incidence of anaemia, and as a consequence, will improve Haitian children's school attendance and educational outcomes.

As yet, the full impact of Nutriset's locally oriented approach is unknown.[64] That said, Nutriset is quick to declare the success of its approach by publishing

---

63 See, for example, Collins, S. 2001. Changing the way we address severe malnutrition during famine. *Lancet*, 358(9280), 498-501; Briend, A., Lacsala, R., Prudhon, C., Mounier, B., Grellety, Y. and Golden M.H. 1999. Ready-to-use therapeutic food for treatment of marasmus. *Lancet*, 353(9166), 1767-1768.

64 In 2011, Plantéte D'Entrepreneurs, a French non-governmental organization, conducted a social impact assessment of the PlumpyField Network at Hilina Enriched Foods in Ethiopia. The Plantéte D'Entrepreneurs examined how the PlumpField network affected the transfer of skills, capacity and know-how, imported technology, the local economy, job creation and standards of living. Unfortunately the results of the Plantéte D'Entrepreneurs study are not yet publicly available. It also appears that the study was commissioned by Nutriset. Personal communication with Hilina Enriched Foods, 22 November 2012.

various figures on employment (over 500 jobs in developing countries) and benefits to local farmers (with, as already noted, over 800 tons of peanuts; over 870 tons of edible oils; and over 970 tons of sugar purchased from local farmers).[65] PlumpyField members also proclaim the success of the PlumpyField network. The first local factory was opened in Malawi in 2005 and since then a number of PlumpyField members have shown positive signs from their involvement in the PlumpyField network. Established in 1998 to manufacture a range of food products specifically designed to alleviate malnutrition and other micronutrient deficiencies in Ethiopia, Hilina Enriched Foods became a PlumpyField member in 2006. In this time, and through the use of Nutriset's patented products (particularly Plumpy'Nut) Hilina claims that '70% of the raw materials required for the production of Plumpy nut are produced locally' and that this 'translates into many positive spin-off effects for the local economy'.[66] Since becoming a PlumpyField member, Hilina has opened a new distribution centre, has increased its production capacity and become the first Ethiopian company to gain the ISO 22000: 2005 food safety certification by demonstrating that they have the ability to control food safety hazards in order to ensure that their food and nutritional products are safe.[67] Furthermore, UNICEF signed a long-term agreement with Hilina for the supply of Plumpy'Nut.[68] Another local manufacturer, Haitian-based Meds & Food for Kids (whose mission is to improve access to innovative nutritional solutions meeting international quality standards through local production, close to the populations requiring them), has also benefited from becoming a PlumpyField member. Recently, on 26 October 2012, Meds & Food for Kids opened a new 18,000 square foot facility in Cap Haiten, Haiti. Prior to this Meds & Food for Kids operated out of the second floor of a converted home. Significantly, the new factory employed up to 60 people during the construction process and has the capacity to produce ten times more of the Meds & Foods for Kids' version of Plumpy'nut, known as Medika Mamba. Meds & Food for Kids employ 22 people to run the factory and three more in agricultural programs to boost agricultural capacity and who source peanuts from 300 local farmers.[69] Finally, as we saw earlier, Meds & Foods for Kids have been awarded a competitive United States Department of Agriculture grant to develop and implement a school snack program through Haiti's National School Lunch Program.

65 *Key Figures*, above n 49.

66 Hilina Enriched Foods. 2012. [Online]. Available at: http://www.hilinafoods.com/about_hilina.php [accessed: 22 November 2012].

67 ISO. 2012. ISO 22000: 2005. *Food Safety Management Systems-Requirements for Any Organization in the Food Chain*. An overview is available at: https://www.iso.org/obp/ui/#iso:std:iso:22000:ed-1:v1:en [accessed: 4 February 2013].

68 Hilina Enriched Foods, above n 66.

69 Meds & Food for Kids. 2012. *You're Invited to Celebrate with MFK in Haiti!* [Online]. Available at: http://mfkhaiti.org/index.php/mfk-blog/2012/9/5/youre-invited-to-celebrate-with-mfk-in-haiti1/ [accessed: 4 February 2013].

Concluding Thoughts … Thinking about Intellectual Property and Food Sociologically

This chapter began with the question of whether intellectual property law can help feed the world. In the preceding pages I have explored some of the assumptions that underpin the justification of extending intellectual property law to agricultural research and food production. While it is an exaggeration to say that technology is not important in achieving food security, when thinking about the role that intellectual property plays in food security it is vital to not single-and-narrow-mindedly associate intellectual property with stimulating agricultural research and food production. Doing so, I have argued, results in a *technology trap*; a predicament in which intellectual property specifies food security as technology at the expense of the relations in which food is produced, accessed and distributed.

In the context of food insecurity, technology-based justifications are complicated by the fact that food insecurity is a physical, economic, political and social problem. More to the point: food insecurity is not about food *per se* but about the relations within which food is produced, distributed and accessed. Informed by a sociological imagination, this chapter has shown how intellectual property is being used by a French company, Nutriset, as a vehicle for social change to encourage and support the local production of their food and nutritional products. While more needs to be done to assess intellectual property's role in feeding the world, the PlumpyField supply network suggests that intellectual property can play a pivotal role in encouraging and supporting local participation in the production and distribution of food and nutritional products. Indeed, the PlumpyField network, and its associated patents and trade marks, allow Nutriset to control the circumstances in which their products, processes and names and logos are used. In this way, Nutriset's intellectual property is being used to create a local presence, help build local capacity, facilitate access to food and nutritional products, and ensure the quality of food and nutritional available in developing countries.

Finally, a word of warning. This chapter should not be read as promoting, in all situations, the use of intellectual property over agricultural research and food production. Using intellectual property to help alleviate hunger, malnutrition and food insecurity is not a one-size-fits-all solution as results produced by intellectual property are contingent upon the relations between law, science, technology, politics, attitudes, practices and individuals. To answer my original question – can intellectual property law can help feed the world? – with anything but caution, circumspection and qualification (over)-simplifies a complex problem. This means that the question of whether intellectual property law can play a meaningful role in alleviating hunger, malnutrition and food insecurity must be answered on a case-by-case basis, taking into account the particular mixture of physical, economic, political and social problems facing those people affected. This, according to C.W. Mills, necessitates a 'capacity to shift from one perspective to another, and in the process to build up an adequate view of a total society and its components'.[70] Only then can we begin to answer the question of whether intellectual property can help feed the world.

Appendix

Usage Agreement No. USAGE NUMBER

This Agreement is concluded by and between:

BENEFICIARY, a TYPE of BUSINESS, registered IN COUNTRY under number REGISTRATION NUMBER, with its registered headquarters at HEAD OFFICES, COUNTRY, and represented by TITLE FIRST NAME SURNAME, acting in HIS/HER capacity as JOB POSITION, duly designated for this purpose,

Hereafter referred to as the " Beneficiary ",

On the one hand,

AND

NUTRISET S.A.S., Société par actions simplifiée, registered with the "Registre du Commerce et des Sociétés" (Trade Register) of Rouen, France, under number B 337 986 798, with its registered headquarters at Le Bois Ricard, BP 35, 76770 Malaunay, France, and represented by Mrs. Adeline Lescanne-Gautier, acting in her capacity as General Manager, duly designated for this purpose,

Hereafter referred to as " Nutriset ",

On the other hand,

Hereafter referred to individually as the "Party" or jointly as the "Parties", depending on the context.

WHEREAS THE FOLLOWING HAS BEEN PRESENTED :
Nutriset is entirely dedicated to the research, development, production, marketing and distribution of products to improve the nutritional status of children and other vulnerable persons in developing countries.

In 1996, Nutriset and the IRD (French public Institute for Research and Development) invented and patented a technological process consisting of a method for preparing and using food supplements which enabled the development of the product Plumpy'nut®, an alternative to products used until then for the treatment of severe malnutrition.

The product Plumpy'nut® and other products of this type since developed by Nutriset, are widely used in humanitarian aid programs for the prevention and treatment of different forms of acute malnutrition affecting vulnerable groups in developing countries.

In an effort to facilitate access to and the availability of this type of product for populations suffering from the affliction of malnutrition, Nutriset, in agreement with the IRD, has initiated and developed a network of producers united around a common project: the fight against malnutrition. This network, named PlumpyField, has more than ten members in 2011, producing, marketing and distributing Nutriset products using Nutriset know-how.

Nutriset and the IRD wish to go even further in their approach to ensuring access to products for the fight against malnutrition. It is in this spirit that the two partners have wished to allow companies in developing nations to benefit from their common patents and, for this purpose, enter into Usage Agreements via a simplified internet-based system.

70 Mills, n 1 above. 211.

In accordance with an agreement between the IRD and Nutriset, Nutriset is the exclusive user of their jointly filed patents and is fully authorized to conclude Usage Agreements.

BENEFICIARY, whose main business is: SECTORS OF ACTIVITY IN COUNTRY, wishes to contribute to the field of nutrition by producing nutritional products for consumption by vulnerable populations.

To this end, BENEFICIARY and Nutriset wish to enter into a non-exclusive and non-transferable Patents Usage Agreement, according to the terms and conditions defined in the present Agreement.

IN CONSEQUENCE, the Parties agree as follows:

Article 1 : Definitions

Each of the words and expressions mentioned below in the present Agreement will have the following meanings:

« **Agreement** » : refers to the present Usage Agreement and its appendixes.

« **Patents** » : refers to patents jointly held by Nutriset and the IRD in the Developing countries stipulated in Appendix A of the present Agreement.

« **Products** » : refers to the nutritional products resulting from developments of the Beneficiary under the terms of the present Agreement and entering into the scope of the Patents.

« **Developing Countries** » : refers to the countries listed as developing countries according to the HPI -1 (Human Poverty Index) classification of the United Nations Development Program (UNDP) with the exception of Niger and Mozambique, with which exclusive agreements have been concluded by Nutriset.

« **Contractual Year** » : designates each period of 366 consecutive days, starting from the first day when the present Agreement enters into effect until the last day of this period.

« **Net turnover** » : refers to the amount of Products sales invoiced by the Beneficiary during the Contractual Year, excluding taxes, discounts, rebates, reductions, returns and transportation costs.

« **Territory** » : refers to the Developing Countries in which Nutriset and the IRD are joint holders of the Patents.

« **Working-days** » : means the days really worked within Nutriset, except Saturdays and Sundays, bank holidays and Nutriset closing days.

Article 2 : Usage Agreement

2.1. Nutriset grants the Beneficiary, who accepts, a non-exclusive and non-transferable authorization of use of the Patents in the Territory (as defined in Article 3).

2.2. The Beneficiary certifies that it is a private or public profit or non-profit company, non-governmental organization or society which

- is real and legitimate,
- has its production and business site, headquarters and main shareholders (at least 51% of the capital) based IN COUNTRY.

2.3. The Beneficiary undertakes, within the framework of the use of the Patents in the Territory, to develop its own

- Page 2 -

formulae, recipes and Products, without assistance from Nutriset and the IRD. Nutriset will not in any way be held responsible for and/or associated with the Beneficiary's Products, particularly in case of defectiveness of the Products which could cause prejudice to third parties.

2.4. The Beneficiary undertakes to develop and implement its own quality system. Nutriset will not in any way be held responsible for and/or associated with this quality system. The Beneficiary therefore agrees not to claim that its Products follow Nutriset quality assurance and quality control rules.

2.5. The Beneficiary is entirely responsible for obtaining any authorization required to manufacture and use the Products in the Territory.

2.6. The Usage Agreement is personal and non transferable. In consequence, the Beneficiary may in no case transmit the present Agreement to a third party, nor authorize a third party directly or indirectly to use the Patents.

The present Agreement may not be considered as part of the Beneficiary's assets, nor used by an official receiver, nor be used as an asset brought into a business by the Beneficiary.

2.7. The Usage Agreement is non-exclusive. Nutriset therefore reserves the right to grant Patents use authorizations to third parties in the Territory.

2.8. The Beneficiary states and recognizes that the present Agreement does not violate its bylaws or contractual commitments, that it possesses the technical and financial means to execute the present Agreement and that, generally speaking, nothing prevents it from perfectly executing the present Agreement. Particularly, the Beneficiary declares and acknowledges that it is not nor in a situation in which it is liable to Nutriset for notably any liabilities incurred neither in dispute with Nutriset.

Article 3 : Geographic coverage of the Agreement
In the context of the Patents usage authorization granted to it, the Beneficiary undertakes to produce the Products solely IN COUNTRY, for utilization within the Territory.

Article 4 : Compensation
4.1. In compensation for the authorization granted under the terms of the present Agreement, the Beneficiary undertakes to:
- recognize Nutriset and the IRD's contribution to in the Nutrition sector through the development of innovative and high-quality nutritional products.
- to pay a "Contribution to the IRD's research works". This contribution is to be 1% of the Net Turnover earned by the Beneficiary during the Contractual Year. It is understood that Nutriset does not expect any financial compensation under the terms of execution of the present Agreement.

4.2. Terms of payment of the Contribution. The Contribution due for the Contractual Year shall be paid directly to the IRD, within a maximum of thirty (30) days from the anniversary date of the present Agreement.

On the anniversary date of the present Agreement, the Beneficiary undertakes to complete an "online declaration form" stating its annual "Contribution to the IRD's research works". This web-based form is available at Nutriset's internet site via the interface "My Space – Usage Agreement by simplified access".

The Beneficiary undertakes to keep as detailed accounts as possible of all Product sales, so as to enable the IRD to confirm the accuracy of the "Contribution to the IRD's research works" due from the Beneficiary.

Upon the request of the IRD, the Beneficiary agrees to permit an independent Certified Public Accountant to be appointed to examine said accounts at the Beneficiary's offices, with reasonable notice and at reasonable times, solely to check the calculation of the "Contribution to the IRD's research works" due by the Beneficiary. The IRD will bear the costs of these verifications within the limits of reasonable costs and fees.

Article 5 : Names, Brands and distinctive signs

5.1. The Products manufactured, marketed and distributed by the Beneficiary, its packaging and means of communication must in no way, shape or form reproduce and/or imitate the brands, logos, packaging, graphic identity or any other distinctive signs used by Nutriset for its own products, packaging and communication media.

5.2. The Beneficiary undertakes to not register or have registered either directly or indirectly any brands, logos or distinctive signs identical and/or similar to those owned by Nutriset in the Territory or anywhere else in the world.

5.3. The Beneficiary undertakes to develop and use its own brands, logos, packaging and graphic identity for the marketing, promotion and distribution of its Products in the Territory.

5.4. The Beneficiary undertakes to clearly and legibly label "Manufactured under Usage Agreement No USAGE NUMBER" on all primary and secondary packaging of the manufactured Products.

Article 6 : Communication, promotion and information

6.1. The Beneficiary undertakes in all its communications, promotions and information to:
- provide information which reflects the reality of the present Agreement,
- communicate in its capacity as Beneficiary without ever evoking or seeking assimilation with Nutriset and/or Nutriset's partners within the network of local producers (the PlumpyField network) nor seeking or evoking any label of scientific approval from the IRD,
- nor associate its Products in any way whatsoever with those of Nutriset or of the PlumpyField network,
- nor refer to Nutriset products or brands in any way, shape or form whatsoever,
- nor make any defamatory and/or denigrating remarks about this Agreement, Nutriset, the PlumpyField network or the IRD.

The Beneficiary further undertakes to clearly indicate in any commercial quotation or reply to a call for bids concerning the Products that it is authorized under a Usage Agreement with Nutriset.

6.2. Nutriset undertakes in all its communications to provide information which reflects the reality of the present Agreement.

Article 7 : Intellectual property rights and counterfeiting

7.1. Nutriset certifies that the Patents jointly held with the IRD in the Territory are in force.

7.2. Nutriset reserves all Patents rights in the Territory. To this end, Nutriset will be solely responsible for any procedures concerning the granting or defense or maintenance in force of the Patents. In case of nonpayment of a Patent annuity in the Territory, Nutriset will inform the Beneficiary. The Parties agree that nonpayment of a Patent annuity is not a reason for terminating the Agreement.

7.3. The Beneficiary recognizes Nutriset's Patents rights and undertakes, for the duration of the present Agreement and thereafter not to directly or indirectly dispute Nutriset's Patents rights nor question their validity.

7.4. Should the Beneficiary becomes aware of any act of counterfeiting or unauthorized use of the Patents by a third party in the Territory, the Beneficiary will inform Nutriset as quickly as possible.

Article 8 : Guaranties

The Beneficiary assumes any and all risks associated with the present Agreement. Nutriset makes no guarantee in any way, shape or form whatsoever with regard to the utilization of the Patents and shall not be held liable in any way to the Beneficiary concerning but not limited to the design, manufacture, safety, quality, efficacy or use of the Products.

Article 9 : Waiver of recourse

9.1. Should a third party take legal action against the Beneficiary with regard to the use of the Patents, the Beneficiary undertakes not to take action against Nutriset but to immediately inform the latter of said action and provide it with a copy. Nutriset undertakes to provide the Beneficiary with any data, record or proof (including Patent ownership), testimony or cooperation reasonably necessary to facilitate the Beneficiary's defense.

9.2. The Beneficiary definitively and irrevocably waives any recourse, legal proceedings or extrajudicial action, as well as any claim against Nutriset and the IRD, regarding Patents and the present Agreement.

Article 10 : Relations between the Parties

No terms of this Agreement may be construed or reputed to consider any partnership, mandate, employer-employee relation or association between the Parties.

Article 11 : Severability clause

Should one or more clauses of this Agreement be declared invalid, illegal or inapplicable for any reason, by a court or by any authority having due jurisdiction, the clause or clauses shall be amended by Nutriset to fit the Parties' intentions as closely as possible; nevertheless, the rest of the Agreement shall not be affected.

Article 12 : Duration

This Agreement shall enter into force within a maximum of seven (7) Working-days from the date of receipt by Nutriset (Date of entry into force) of two originals of the present Agreement signed and initialed on each page by the Beneficiary, and shall be valid in the Territory for three (3) years, tacitly renewable, until the date of expiration of the last Patent in the Territory.

On expiration of this three-year (3yr) period, unless one of the Parties has given notice of cancelation by registered express mail with return receipt respecting a notice period of at least six (6) months, this Agreement shall be renewed for successive three-year (3yr) periods, unless notice has been given six (6) months prior to the expiration of each period as stated above, until expiration of the last Patent in the Territory.

Article 13 : Termination

This Agreement may be terminated under the following conditions:

- In case of non-execution by one or the other of the Parties of one or more significant obligations incumbent upon them in virtue of the present Agreement, the adversely affected Party shall send a formal letter by registered express mail with return receipt to the other Party specifically outlining where or how the Party has failed in its obligations. The present Agreement shall then be rightfully terminated in a period of thirty (30) days from reception of said letter if the receiving Party has not remedied the cited failure.
- In case of liquidation, cessation of business or change of ownership of the Beneficiary,
- By mutual written consent of the Parties.

Termination of this Agreement for any reason shall not affect any rights, obligations and responsibilities accumulated up to the date of termination of this Agreement.

Termination of this Agreement either partially or totally shall not free the Beneficiary of its obligations to pay any

"Contribution to the IRD's research works" previously due to the IRD at the effective date of termination

In case of termination, the Beneficiary shall stop all production, sales and distribution of the Products in the Territory. Nonetheless, Nutriset will allow the Beneficiary to continue to sell and distribute remaining stocks of the Products (but not to manufacture the Products) for a period of six (6) months from the date of termination of this Agreement. During this period, the Beneficiary will continue to sell and distribute the Products in accordance with the terms and conditions of this Agreement and pay the IRD the corresponding "Contribution to the IRD's research works".

Article 14 : Communication of the Agreement

14.1. The Beneficiary accepts Nutriset's right to communicate the Beneficiary's contact details publically on the Nutriset website on any other medium.

14.2. Nutriset will remove the Beneficiary's contact details from the Nutriset website in case of termination of this Agreement and may give information on this in its communications.

Article 15 : Applicable law

The present Agreement is governed by the French laws and regulations.

Article 16 : Settlement of dispute

The Parties agree to bring any disputes that cannot be solved amicably with regard to interpretation, validity, execution or termination of the present Agreement to the competent courts of Paris, France.

EXCLUSION OF GUARANTEE: unless otherwise explicitly stated in the Present Agreement, the Parties expressly agree that they decline any express or implicit guarantee including, without limitation, commercial value, suitability for a specific use, and non-counterfeiting.

IN WITNESS WHEREOF, the Beneficiary and Nutriset execute the present Agreement, by their respective duly designated representatives, as from the Date of entry into force of the present Agreement.

Date of receipt of this Agreement by Nutriset :

BENEFICIARY	**NUTRISET S.A.S**
By:	By:
Name: FIRST NAME SURNAME	Name: Adeline Lescanne-Gautier
Title: JOB POSITION	Title: General Manager

Appendix A: Developing countries in which Nutriset and the l'IRD held Patents

Patent 1

Title	Country	Filling Date	Patent Application No.	Date of Patent	Patent No.	Date of Expiry
Food or nutritional supplement, preparation method and uses	Ghana	19/11/1998	AP P 00 01827	06/10/2003	AP 1211	19/11/2018
	Gambia	19/11/1998	AP P 00 01827	06/10/2003	AP 1211	19/11/2018
	Kenya	19/11/1998	AP P 00 01827	06/10/2003	AP 1211	19/11/2018
	Lesotho	19/11/1998	AP P 00 01827	06/10/2003	AP 1211	19/11/2018
	Malawi	19/11/1998	AP P 00 01827	06/10/2003	AP 1211	19/11/2018
	Sudan	19/11/1998	AP P 00 01827	06/10/2003	AP 1211	19/11/2018
	Swaziland	19/11/1998	AP P 00 01827	06/10/2003	AP 1211	19/11/2018
	Uganda	19/11/1998	AP P 00 01827	06/10/2003	AP 1211	19/11/2018
	Zimbabwe	19/11/1998	AP P 00 01827	06/10/2003	AP 1211	19/11/2018
Aliment ou supplément nutritionnel, son procédé de préparation et ses utilisations	Bénin	19/11/1998	1 200 000 152	29/12/2000	11416	19/11/2018
	Burkina Faso	19/11/1998	1 200 000 152	29/12/2000	11416	19/11/2018
	Cameroun	19/11/1998	1 200 000 152	29/12/2000	11416	19/11/2018
	Centre Afrique	19/11/1998	1 200 000 152	29/12/2000	11416	19/11/2018
	Congo	19/11/1998	1 200 000 152	29/12/2000	11416	19/11/2018
	Côte d'Ivoire	19/11/1998	1 200 000 152	29/12/2000	11416	19/11/2018
	Gabon	19/11/1998	1 200 000 152	29/12/2000	11416	19/11/2018
	Guinée	19/11/1998	1 200 000 152	29/12/2000	11416	19/11/2018
	Guinée Bissau	19/11/1998	1 200 000 152	29/12/2000	11416	19/11/2018
	Guinée Equatoriale	19/11/1998	1 200 000 152	29/12/2000	11416	19/11/2018
	Mali	19/11/1998	1 200 000 152	29/12/2000	11416	19/11/2018
	Mauritanie	19/11/1998	1 200 000 152	29/12/2000	11416	19/11/2018
	~~Niger~~	~~19/11/1998~~	~~1 200 000 152~~	~~29/12/2000~~	~~11416~~	~~19/11/2018~~
	Sénégal	19/11/1998	1 200 000 152	29/12/2000	11416	19/11/2018
	Tchad	19/11/1998	1 200 000 152	29/12/2000	11416	19/11/2018
	Togo	19/11/1998	1 200 000 152	29/12/2000	11416	19/11/2018

Patent 2

Title	Country	Filling Date	Patent Application No.	Date of Patent	Patent No.	Date of Expiry
High energy complete food or nutritional supplement, method for preparing same and uses thereof	Gambia	23/10/2001	AP/P/2003/002789	31/07/2006	AP 1647	23/10/2021
	Ghan	23/10/2001	AP/P/2003/002789	31/07/2006	AP 1647	23/10/2021
	Kenya	23/10/2001	AP/P/2003/002789	31/07/2006	AP 1647	23/10/2021
	Lesotho	23/10/2001	AP/P/2003/002789	31/07/2006	AP 1647	23/10/2021
	Malawi	23/10/2001	AP/P/2003/002789	31/07/2006	AP 1647	23/10/2021
	~~Mozambique~~	~~23/10/2001~~	~~AP/P/2003/002789~~	~~31/07/2006~~	~~AP 1647~~	~~23/10/2021~~
	Sierra Leone	23/10/2001	AP/P/2003/002789	31/07/2006	AP 1647	23/10/2021
	Sudan	23/10/2001	AP/P/2003/002789	31/07/2006	AP 1647	23/10/2021
	Swaziland	23/10/2001	AP/P/2003/002789	31/07/2006	AP 1647	23/10/2021
	Tanzania	23/10/2001	AP/P/2003/002789	31/07/2006	AP 1647	23/10/2021
	Uganda	23/10/2001	AP/P/2003/002789	31/07/2006	AP 1647	23/10/2021
	Zimbabwe	23/10/2001	AP/P/2003/002789	31/07/2006	AP 1647	23/10/2021
Aliment complet ou supplément nutritionnel à haute valeur énergétique, son procédé de préparation et ses utilisations	Bénin	23/10/2001	1 2003 00124	17/09/2004	12 530	23/10/2021
	Burkina Faso	23/10/2001	1 2003 00124	17/09/2004	12 530	23/10/2021
	Cameroun	23/10/2001	1 2003 00124	17/09/2004	12 530	23/10/2021
	Centre Afrique	23/10/2001	1 2003 00124	17/09/2004	12 530	23/10/2021
	Congo	23/10/2001	1 2003 00124	17/09/2004	12 530	23/10/2021
	Côte d'Ivoire	23/10/2001	1 2003 00124	17/09/2004	12 530	23/10/2021
	Gabon	23/10/2001	1 2003 00124	17/09/2004	12 530	23/10/2021
	Guinée	23/10/2001	1 2003 00124	17/09/2004	12 530	23/10/2021
	Guinée Bissau	23/10/2001	1 2003 00124	17/09/2004	12 530	23/10/2021
	Guinée Equatoriale	23/10/2001	1 2003 00124	17/09/2004	12 530	23/10/2021
	Mali	23/10/2001	1 2003 00124	17/09/2004	12 530	23/10/2021
	Mauritanie	23/10/2001	1 2003 00124	17/09/2004	12 530	23/10/2021
	~~Niger~~	~~23/10/2001~~	~~1 2003 00124~~	~~17/09/2004~~	~~12 530~~	~~23/10/2021~~
	Sénégal	23/10/2001	1 2003 00124	17/09/2004	12 530	23/10/2021
	Tchad	23/10/2001	1 2003 00124	17/09/2004	12 530	23/10/2021
	Togo	23/10/2001	1 2003 00124	17/09/2004	12 530	23/10/2021

Chapter 9

Geographical Indications and Agricultural Community Development: Is the European Model Appropriate for Developing Countries?

Graham Dutfield[1]

Introduction

Are geographical indications (GIs) truly beneficial for developing countries? And, if they are, how can they best be used to promote agricultural community development? This chapter seeks answers, taking into account European experience as well as developing country experience to the extent that the latter is able so far to provide any clear lessons.

According to the World Trade Organization's *Agreement on Trade-related Aspects of Intellectual Property Rights* (TRIPS), GIs are 'indications which identify a good as originating in the territory of a Member, or a region or locality in that territory, where a given quality, reputation, or other characteristic of the good is essentially attributable to its geographical origin'.[2] The idea of communicating origin to consumers and other traders by words or signs goes far back in time. But with the inclusion of GIs in TRIPS, they are now part of international trade law.

Unquestionably, the underlying purpose of including GIs in the text of TRIPS was to protect the interests of European producers, mostly those of wines and spirits who are able to enjoy additional protection as compared to other types of product that may be important items of trade within Europe, but are less important globally. These include foodstuffs such as cheeses and processed meats. Little thought was given during the negotiations to the use of GIs as means to further the

1 This chapter benefits considerably from earlier research I did on GIs that was funded by International Institute for Environment and Development (IIED) and the defunct TradeCom Facility. Accordingly, I wish to express my gratitude to these organizations and to the following individuals: Emma Blackmore and Krystyna Swiderska (IIED), Lingston Cumberbatch (Tradecom), Chinwe Isiadinso (African Centre for Study, Development, Research and Civic Education (ACCE)), David Jordan and Shirnaya Stephen (government of Saint Lucia), and Bernard O'Connor (my collaborator on the Saint Lucia project).

2 *Agreement on Trade-related Aspects of Intellectual Property Rights*, Art 22.1.

trade and development interests of the developing countries, even less to promote, valorize and generate income from traditional knowledge-based products.

However, since 1995 when TRIPS came into force, there has been a growth of interest in the trade and development aspects of GIs. Even so, few detailed empirical studies have been conducted on the operations of GI regimes in developing countries, or that have any obviously direct applicability to such nations.

Many European advocates of GIs assert that many relatively backward agricultural communities *have* gained from use of GIs. In consequence, their wider adoption could benefit such populations elsewhere in the world. The European Union, as does the Swiss government, funds technical assistance programmes that promote the adoption of GI laws and regulations, and guidance on how to implement them. Putting aside the obvious political motivations behind such efforts, one must admit that there is a genuine conviction among officials at the European Commission that GIs are good for developing countries, including in the context of local development.

Most developing countries *do* see a value in GIs, and more and more countries are passing *sui generis* GIs legislation with TRIPS and the EU system forming the key points of reference whether or not newly framed national regimes look much like it. Further, they generally agree with the European Union's position on extending the additional protection available under TRIPS for wines and spirits to all other types of product. But beyond that the consensus breaks down.

Some developing countries are understandably concerned that since most of the world's GIs, especially those with well-established market positions, are European, increased intellectual property rents will in the short to medium term accrue to Europe whilst sales of products of theirs that compete with GI-protected European ones will be harmed. Indeed, those countries established in their modern form by European settlers tend to be the most sceptical about the worth of GIs and the most cynical about European motivations. It is not difficult to see why: through a process described by Alfred Crosby, which he called 'ecological imperialism',[3] settlers turned vast tracts of land including indigenous peoples' territories into what one might call 'offshore Europes' containing plants and animals carried with them from Europe that were cultivated and improved with European agricultural practices, and in sites and regions having the same place names as those from the continent they had left. In this context, one should not be too surprised that it was Australia and the United States, both major producers of old world agricultural commodities, foodstuffs and beverages, who made a formal complaint at the WTO against the European Community for its methods of protecting GIs, largely, it must be said, its own. Both countries alleged that these were incompatible with TRIPS concerning national treatment, most-favoured-nation and trademark protection. The Dispute Panel formed to consider the merits of the complaint did not fully

3 Crosby, A. 1993. *Ecological Imperialism: The Biological Expansion of Europe, 900-1900.* Cambridge: Cambridge University Press.

accept the position of Australia and the US, but did concur with the complainants' view that the Regulation violated the national treatment provision in TRIPS by discriminating against foreign nationals seeking to have their GIs registered in Europe.[4] The overwhelming majority of indications in force in Europe continue to be for products of European origin, although there has been a recent increase in non-European registrations.

Be that as it may, there is no disputing that Europe has by far the most experience and could offer useful learning for developing countries. But it is important to bear in mind the very different legal, economic, social and political contexts of different countries, as well as the capacity challenges as demonstrated by the Saint Lucia case study described below. There are some well-known and long-established GIs in developing countries that have all been quite successful, albeit usually protected by trademarks rather than *sui generis* GI regimes. These include Darjeeling tea from India, Jamaica Blue Mountain coffee, tequila from Mexico, and coffees from Colombia and Guatemala. However, these are prestigious products, with goodwill and profitable exports established long before their GI protection.

For small-scale and widely dispersed producers, such as coffee growers in Ethiopia to give just one example, boundary-setting, standardization and quality control measures can be very difficult, if not impossible, to put into place. Regarding the lessons learned, as one commentator has observed, 'although substantial research has been conducted on European GI systems …, very few comprehensive studies have been carried out on the specific challenges associated with the implementation of GI schemes outside of Europe'.[5]

A Very European IP? The European *sui generis* Geographical Indications Regimes

The TRIPS definition says little about what a GIs regime should actually look like. If we view GIs as a category of IP, they would appear to have a very short history. Related terms like indications of source and appellations of origin were included in the 1891 *Madrid Agreement on the Repression of False or Deceptive Indications of Source in Goods*, and in the 1958 *Lisbon Agreement for the Protection of Appellations of Origin and their International Protection*. However,

---

4 World Trade Organization. 2005. *European Communities – Protection of Trademarks and Geographical Indications for Agricultural Products and Foodstuffs, WT/DS174/R*. Geneva: World Trade Organization. For a detailed discussion on this case see Evans, G.E. and Blakeney, M. 2006. The protection of geographical indications after Doha: Quo vadis? *Journal of International Economic Law*, 9(3), 575-614.

5 Bowen, S. 2010. Development from within? The Potential for geographical indications in the global South. *Journal of World Intellectual Property*, 13(2), 231-252, 242.

'geographical indication', of which indications of source and appellations of origin may be considered a subset, is much more recent.[6]

More than 15 years after TRIPS came into force the legal concept of GIs remains somewhat fuzzy. How WTO members should provide the legal means to protect such indications is 'strategically vague' allowing for much flexibility. Accordingly, there are diverse ways by which an indication of geographical origin, and the product so indicated, may be protected legally. This gives WTO members broad scope to draw up implementing laws and regulations as they see fit. Different countries may well protect them under original and specific (*sui generis*) GI laws, as in the European Union and India. But they may alternatively – *or additionally* – be protected largely or completely under the following: an appellation (or designation) of origin regime; trademark law, or in some countries influenced by English law under the common law tort of passing off as with unregistered trademarks; or indirectly through unfair competition law.

They may in fact be embedded within areas of law and regulation other than IP according to the underlying purpose of protecting them, such as: consumer protection rules relating to trade descriptions or food product labelling; cultural heritage regulations and policies; or rural development regulations and policies.

The role of the state may be minimal, simply providing the legal framework and granting rights. Or else the state may be directly involved in overseeing the regulation and enforcement of GIs. Thus, GIs are not a type of IP as is the patent, copyright or trade mark. Perhaps it is best to think of 'geographical indication' as a convenient catch-all term for a variety of mostly currently-existing laws and regulations concerning products that are differentiated by possessing features or a reputation attributable to their spatial origins. In this sense, GIs have a much longer history than the term itself.

What we can say with certainty is that like *sui generis* plant variety protection (and its international incarnation as UPOV), GIs are of European origin. Indeed, it is no secret that the reason for their inclusion in the Uruguay Round of trade negotiations of the late 1980s and early 1990s that lead to the establishment of the WTO was a direct consequence of proposals made by the European Community and by Switzerland. More GIs relate to goods of European origin than to those of any other region or country in the world, in fact most likely the rest of the world put together. While important to note this, it does not automatically follow that GIs are unsuited to other regions that differ markedly from Europe in terms of culture and economic conditions. But at the very least, it might help to explain the

---

6 The phrase 'geographical indications of origin' was used by Robin Jacob, subsequently a British Court of Appeal judge in an article published in 1980. But its common usage nowadays is obviously due to TRIPS and to more recent EU law. Jacob, R. 1980. The Protection of Geographical Indications of Origin in the United Kingdom, in *Protection of Geographic Denominations of Goods and Services*, edited by H.C. Jehoram. Alphen aan den Rijn & Germantown: Sijthoff & Noordhoff, 135-148.

continued relative absence of effective national GI regimes in more than a handful of developing countries.

Probably the longest established geographical indications regime is the French appellations of origin system for products considered to be distinctively local due to a combination of traditional know-how and highly localized natural conditions.[7] This dates back to a law passed in 1905 but it became truly effective only from the mid-1930s once policymakers learned how best to remedy the original system's flaws. The appellations of origin system evolved in response to problems of fraudulent or misleading labelling, poor quality control and overproduction. A government agency, the Institut National de l'Origine et de la Qualité (INAO) validates *Appellations d'Origine Contrôlée*, so that producers of wines, spirits, dairy and agricultural products, whose goods are renowned for their distinctive qualities and geographic origins, are protected from those who would undermine or exploit their good reputation by making similar, but false, claims. Applications are made by producer associations, and these are scrutinized by committees of inquiry composed of members of the national committee and expert appointees who may draw up production conditions that users of the indication must comply with. The INAO, the committees, and the associations cooperate to make the system of accreditation, management and enforcement, operate effectively.[8] There are similar regimes in Italy, Spain and Switzerland among other European countries. Europe has the most elaborate GI regime in the world, covering wines,[9] spirits,[10] agriculture and foodstuffs. For the latter two types of product, which are the most relevant ones to the discussion here, the European Union currently has two legal instruments in effect:

1. Council Regulation (EC) No 510/2006 of 20 March 2006 on the protection of geographical indications and designations of origin for agricultural products and foodstuffs.

7 Gangjee, D. 2012. *Relocating the Law of Geographical Indications*. Cambridge: Cambridge University Press, 77-115.

8 Roncin, F. 2005. The Protection of Geographical Indication, in *Biodiversity and Local Ecological Knowledge in France*, edited by L. Bérard et al. Paris: CIRAD, IDDRI, IFB and INRA, 175-180.

9 Commission Regulation (EU) No 401/2010 of 7 May 2010 amending and correcting Regulation (EC) No 607/2009 laying down certain detailed rules for the implementation of Council Regulation (EC) No 479/2008 as regards protected designations of origin and geographical indications, traditional terms, labelling and presentation of certain wine sector products.

10 Regulation (EC) No 110/2008 of the European Parliament and of the Council of 15 January 2008 on the definition, description, presentation, labelling and the protection of geographical indications of spirit drinks and repealing Council Regulation (EEC) No 1576/89.

2. Council Regulation (EC) No 509/2006 of 20 March 2006 on agricultural products and foodstuffs as traditional specialities guaranteed.

Under these Regulations, Europe has three types of registered GI. These are protected designations of origin (PDOs), protected geographical indications (PGIs), and traditional specialities guaranteed (TSGs). It is worthwhile to look at these in some detail because despite Europe's relative wealth and longer experience in the use of GIs, their objectives appear on the face of it to be consistent with those that small-scale producers in developing countries might also consider relevant and important. Thus, the preamble of Regulation 510/2006 notes that: 'The promotion of products having certain characteristics can be of considerable benefit to the rural economy, particularly in less-favoured or remote areas, by improving the incomes of farmers and by retaining the rural population in these areas.'

Article 2 of the same Regulation provides definitions of PDOs and PGIs. A PDO is:

> the name of a region, a specific place or, in exceptional cases, a country, used to describe an agricultural product or a foodstuff:
>
> —originating in that region, specific place or country,
>
> —the quality or characteristics of which are essentially or exclusively due to a particular geographical environment with its inherent natural and human factors, and
>
> —the production, processing and preparation of which take place in the defined geographical area;

A PGI means:

> the name of a region, a specific place or, in exceptional cases, a country, used to describe an agricultural product or a foodstuff:
>
> —originating in that region, specific place or country, and
>
> —which possesses a specific quality, reputation or other characteristics attributable to that geographical origin, and
>
> —the production and/or processing and/or preparation of which take place in the defined geographical area.

The definitions are similar, but we can see that the requirements for a PDO are somewhat more stringent in two senses. First, the geographical environment with its natural and human factors is more or less *wholly* responsible for the product or foodstuff's positive features. PGIs also place a lot of emphasis on geography, but it is mere place-origin that counts rather than any special environmental and human factors. Moreover, the responsibility of geography for the specialness of the product is expressed in less absolute terms: 'attributable to' rather than 'essentially or exclusively due to'. Second, PDOs require that production, processing and preparation of the product all take place in the geographical area, while PGIs require that at least one of these be carried out in the area. Like patents and trademarks, PDOs and PGIs must be registered. The Regulation explains the procedures for this.

It is possible to oppose any indication. However, when governments forward applications to the Commission, they tend to be registered whether or not they are particularly rigorously prepared or indeed whether they necessarily stand up to serious critical examination. Some controversial indications have been registered and even survived litigation when many objective observers would not have expected such an outcome (see box).

Box 9.1 PDOs and PGIs in Europe: Some (possibly anomalous) examples

A Protected Designation of Origin: Feta Cheese
Feta is a white cheese stored in brine and made from sheep's milk or a mix of sheep's and goat's milk from local breeds and made using traditional methods. It was registered in 2002 as a cheese made in certain areas of Greece. As is often the case with GIs, the product has for quite some time been produced elsewhere. The question then arises as to whether it is fair to grant a GI when a reasonably objective case could be made for its having become a generic term for damp salty, white sheep or goat milk cheese, and that the public was not misled into thinking that non-Greek feta is really Greek. Nonetheless, in 2005, the Court of Justice of the European Union, perhaps mistakenly, ruled it illegal for any cheese produced outside Greece to be called 'feta', despite the fact that in Denmark, Germany and France locally made cheeses with the word 'feta' in the name had been sold for several decades. Feta is not in fact the name of a place but is the Greek word for 'slice'.*

A Protected Geographical Indication: Cornish Pasty
A pasty is a type of meat and vegetable pie commonly associated with the English county of Cornwall, which is one of the poorest parts of the country suffering from the consequences of the closing down of its mining industry and cultural erosion including loss of its own language. In 2011 'Cornish pasty' was granted PGI status recognizing it as an important aspect of the county's 'culinary heritage', and acknowledging also that (i) the general public tends to associate pasties with Cornwall and (ii) that pasty making does seem to be taken quite seriously there as compared to other parts of England where they were also made. On the other hand, 'Cornish' pasties have been produced outside the county for a very long time and there is no particular reason to suppose that making them in Cornwall and from ingredients sourced within the county results in a product that is different in quality or in any other way from pasties made elsewhere. After all, it is not regarded as

a particularly high-quality product and no special skill is required to make them. Indeed, there is some documentary evidence to suggest that pasties originate not from Cornwall but the neighbouring county of Devon.** Despite this, all existing producers outside Cornwall must either cease their use of the indication or have by request been granted a 3-year phase-out period. They can, of course, continue to call their products pasties; they just have to discontinue using the word 'Cornish'.

* Judgment of the Court (Grand Chamber) of 25 October 2005. Federal Republic of Germany (C-465/02) and Kingdom of Denmark (C-466/02) v Commission of the European Communities. Joined cases C-465/02 and C-466/02. European Court reports 2005, p. I-09115.
** BBC. 2006. Devon Invented the Cornish Pasty. BBC News [Online]. Available at: http://news.bbc.co.uk/1/hi/england/cornwall/6144460.stm [accessed: 9 January 2013].

The Traditional Speciality Guaranteed mark is for products that are traditional in the sense of being used 'on the Community market for a time period showing transmission between generations; this time period should be the one generally ascribed to one human generation, at least 25 years'. It is not necessary that they be produced in a particular area. This feature makes TSGs rather interesting. After all, many indigenous peoples have been forced to inhabit areas other than their original territories yet traditional products continue to be made. Two European examples of TSG products are Traditionally Farmed Gloucestershire Old Spots Pork and Pizza Napoletana. These two names can be used by producers anywhere in Europe: there is no requirement to produce only in Gloucestershire or Naples; but only those products produced according to the specifications laid out by the TSG can actually use the mark. In the former case, the TSG is controlled by the Gloucestershire Old Spots Pig Breeders Club. As for the latter, two associations in the city of Naples applied for the TSG, with three organizations made responsible for verifying compliance with the specification. It is worth noting that the Regulations provide two alternatives: either the name of the product can be used by third parties without authorization from the holders of the TSG; or else use of the name is reserved to producers granted permission from the holder. But in both cases, the TSG sign can only be used by authorized producers.

In describing any type of IP, it is essential to go beyond the subject matter and consider the scope of the rights: what acts by third parties are right holders, permitted users or other interested parties entitled to prevent? Article 13.1 (Protection) of Regulation 510/2006 protects against the following acts:

a. any direct or indirect commercial use of a registered name in respect of products not covered by the registration in so far as those products are comparable to the products registered under that name or in so far as using the name exploits the reputation of the protected name;

b. any misuse, imitation or evocation, even if the true origin of the product is indicated or if the protected name is translated or accompanied by an expression such as 'style', 'type', 'method', 'as produced in', 'imitation' or similar;
c. any other false or misleading indication as to the provenance, origin, nature or essential qualities of the product, on the inner or outer packaging, advertising material or documents relating to the product concerned, and the packing of the product in a container liable to convey a false impression as to its origin;
d. any other practice liable to mislead the consumer as to the true origin of the product.

Similar but less detailed language can be found in Article 17 of Regulation 509/2006 on Traditional Specialities Guaranteed, although it is not necessary that the name itself be protected: exclusivity is only in reference to the TSG symbol or the 'traditional specialities guaranteed' phrase. It depends on whether the producer group concerned has requested that the name of the product be reserved. For this request to be granted, Article 13 stipulates that the name must not already be 'used in a lawful, renowned and economically significant manner for similar agricultural products or foodstuffs'.

It should be clarified that, notwithstanding the feta case suggesting otherwise, when a once-local product becomes generic it can no longer be protected. This has been the fate of such products named after places as Cheddar cheese and Dijon mustard, which can now be produced anywhere. One would suppose that once a product becomes generic there is no going back. And yet reclaiming the indication may not be totally impossible. A partial re-localization has been achieved, for example, with Cheddar cheese. There is now a PDO on 'West Country Farmhouse Cheddar'. However, the permitted area of production is far larger than the village of Cheddar.

Moreover, with political will and sympathetic courts, products that supposedly have become generic in some territories *can* sometimes be re-localized outside and within the jurisdiction. This has been the case for sherry, a fortified wine named by linguistically challenged English people after a district of Spain called Jerez.[11] 'Cyprus sherry', sold under that name since the 1930s, has thus been renamed 'Cyprus fortified wine'. The naming of South African fortified wines as sherry and port had to be phased out under the terms of the 1999 EU–South Africa free trade agreement,[12] according to which all exporting under those names had to cease

11 Similarly, 'blue denim' is a corruption of 'bleu de Nîmes'.

12 Officially: *Agreement on Trade, Development and Cooperation between the European Community and its Member States, of the one part, and the Republic of South Africa, of the other part.*

within eight years, with domestic sale having to end within 12 years.[13] Another good example is the aforementioned feta cheese.

Practical Aspects of GIs in Developing Countries

Evidently, a substantial majority of European Union GIs do come from relatively poor areas of Europe. Given the premium prices achieved by the more successful indications, it may indeed reasonably be asserted that 'GIs have the potential to generate value-added products for farmers in depressed areas'.[14] A frequently cited example is Comte cheese from the French Jura mountain region. When protected as a GI in 1993 'the cheese enjoyed a 20% price differential over Emmenthal produced in Switzerland which is not GI protected. By 2003 this had risen to 46% and production of Comte had risen by 3% a year'.[15]

What of developing countries? The vast majority (about 90 per cent) of GIs are from the OECD countries. Consequently, developing country experiences are quite limited. But this is not to say these countries have been completely inactive. Indeed, Chile protected the grape brandy product 'Pisco' (see below) as early as 1943, requiring the name be reserved 'for an "eau de vie" produced in a certain region of the country'.[16] Several developing countries are parties to the 1958 *Lisbon Agreement for the Protection of Appellations of Origin and their International Registration*. Of these, six have registered appellations.[17] Obviously this is not a large number of countries. In 1999, India passed legislation (see box),[18] motivated in large part by the furore surrounding a notorious US patent relating to basmati rice, as well as trademarks using the 'mati' part of the word, and to concerns about the substantial quantity of tea being falsely labelled as 'Darjeeling'. So far 151 indications have been registered, of which the majority are actually for handicrafts rather than agricultural products. Five rices from the southern state of Kerala are registered; basmati, ironically, is not. A handful of indications are foreign, including 'Champagne' and 'Peruvian Pisco'. While the benefits of the system

---

13 O'Connor and Company. n.d. *Geographical Indications and TRIPs: 10 Years Later ... A Roadmap for EU GI Holders to Get Protection in other WTO Members*, 10-11 [Online]. Available at: http://trade.ec.europa.eu/doclib/docs/2007/june/tradoc_135088.pdf [accessed 9 January 2013].

14 Bowen, above n 5, 234.

15 Watson, J. and Streatfeild, J. 2008. *The Starbucks/Ethiopian Coffee Saga: Geographical Indications as a Linchpin for Development in Developing Countries*, Policy Note 3. Uppsala: Nordic Africa Institute.

16 Aracama Zorraquin, E.D. 1980. The Protection of Geographic Denominations in South America, in *Protection of Geographic Denominations of Goods and Services*, edited by H.C. Jehoram. Alphen aan den Rijn & Germantown: Sijthoff & Noordhoff, 93-96, 94.

17 These are Algeria, Cuba, Mexico, North Korea, Peru and Tunisia.

18 The *Geographical Indications of Goods (Registration and Protection) Act 1999.*

have been uneven, and it is of course still early days, a recent review suggests there have been successes.[19]

Box 9.2 India's Geographical Indications legislation

India was one of the first developing countries to pass a TRIPS-compliant law on geographical indications: The Geographical Indications of Goods (Registration and Protection) Act of 1999. The Act is supplemented by implementing rules adopted in 2002. The Indian government appears to have been motivated to do so by: (i) a notorious United States patent relating to basmati rice, as well as trademarks being granted claiming the 'mati' part of the word in combination with other prefixes; (ii) concerns about the substantial quantity of tea being falsely labelled as Darjeeling; and (iii) a conviction that GIs are suitable for many products in which India has a competitive advantage.

According to this law, an applicant to register a GI may be:
'Any association of persons or producer or any organization or authority established by or under any law for the time being in force representing the interest of the producers of the concerned goods, who are desirous of registering a geographical indication in relation to such goods'. (Chapter III, Section 11(1)).

Despite the complexity of the procedure for acquiring a GI in India, 153 had been registered by July 2011. There are 41 for agricultural products, 97 for handicrafts, 12 for manufactured goods including beverages, and three for foodstuffs. Here are a few examples: Bhalia Wheat, Mahabaleshwar Strawberry, Nashik Grapes (agricultural products); Channapatna dolls and toys, Hand made Carpet of Bhadohi; Mysore Silk (handicrafts); Feni, Mysore Sandalwood Oil, Nashik Valley Wine (manufactures); and Hyderabad Haleem (foodstuff). There are five foreign registrations in India, all of which are foods or beverages: Champagne (France), Napa Valley (USA), Parma Ham (Italy), Peruvian Pisco (Peru) and Scotch Whisky (UK) (Geographical Indications Registry, 2011).

Several developing countries have provided for the registration of GIs in the form of designations or appellations of origin including Brazil, Cuba and Mexico in Latin America, and Algeria and Tunisia in North Africa. These are to a large extent wines, spirits, foodstuffs and agricultural products, and tobacco products. The European Union has given substantial attention to trade and development relations with the more than 70 countries of the African, Caribbean and Pacific (ACP) Group of States. The ACP countries (minus Cuba) and the European Union and its member states signed the *Cotonou Agreement*, a lengthy development cooperation partnership instrument detailing the commitment on the part of each of

19 Das, K. 2010. Prospects and challenges of geographical indications in India. *Journal of World Intellectual Property*, 13(2), 148-201.

the parties 'to work together towards the achievement of the objectives of poverty eradication, sustainable development and the gradual integration of the ACP countries into the world economy'.[20] Chapter 5, on trade-related areas, contains an article on IP that makes explicit reference to GIs. It stresses cooperation, technical assistance and envisages the possibility of engaging in further agreements on protection of GIs and trademarks for certain products.

Jamaica is another country that has been active, though the slowness in implementing legislation is fairly typical. Diplomatically, Jamaica is very pro-GI. It is a supporter of extending additional protection of GIs at the WTO. Up to recently, only certification marks were available to protect Jamaican GIs. However, Jamaica has now opted for a *sui generis* GI law. In 2009, its *Protection of Geographical Indications Act* entered into force along with the implementing regulations. The legislation is closely based on the language of TRIPS Articles 22-24 and accordingly provides a higher level of protection for wines and spirits and a lower level for all other products. The government intends to change that so that the heightened level is available to all products without discrimination. In order to develop a workable infrastructure for the GI system, the government has received technical assistance from the Swiss Federal Intellectual Property Institute.

One of Jamaica's best-known products is its 'Blue Mountain coffee', which is protected in many countries by a certification trademark. It is a high-value product whose quality is directly attributable to local climatic and soil conditions. There are said to be large quantities of coffee being mislabelled as Blue Mountain including in countries where a certification trademark has been granted. Another problem is the sale of blends that include Blue Mountain along with other coffees. There is a strong intention to protect Blue Mountain as a GI under the new law and it is hoped that by doing so it will be easier to take effective legal action in these key overseas markets. Two other products that should be registered soon are 'Jamaica Rum' and 'Jamaica Jerk', a type of seasoning for meats, originally pork but nowadays also poultry and fish. The coffee and rum producers are already organized into associations. This makes it easier for them to take advantage of having use of a GI and taking control over the whole production chain from growing the raw materials to packing or bottling the finished product. The Jerk producers, however, do not have a well-established association. The Bureau of Standards of Jamaica will perform the role of monitoring the production process and maintaining quality.

Nonetheless, it remains the case that few developing country GIs are registered in any part of the world both in absolute and relative terms, with only a small number of them having any indications at all. A number of empirical studies on GIs in developing countries have been conducted that highlight the practical

20 Partnership Agreement between the Members of the African, Caribbean and Pacific Group of States of the one part, and the European Community and its Member States, of the other part, second revision, June 2010, Preamble.

possibilities and challenges, while allowing for comparison with other GIs. It is to these that the discussion now turns.

Spirits of the South: Feni, Tequila and Pisco

Feni is an alcoholic beverage produced in Goa in India and generally associated with that part of the country. In 2009 it was registered as a GI jointly by the Department for Science and Technology of the government of the state of Goa and a producer association called the Goa Cashew Feni Distillers and Bottlers Association. In line with the rules operating in India, the applicants need to submit detailed specifications containing 'historical information about Feni in Goa, details of the materials and techniques of Feni distilling and chemical analysis of the liquor'.[21] Producers and bottlers obviously have a common interest in being about to take advantage of the Feni GI and prevent others whose product does not comply with the specifications but who may wish to pass it off as Feni from doing so. But they are also competitors whose interests may not be identical. Given the stakes involved, it is obviously important that those intending to benefit from the GI are able to agree on the specifications and on any future modifications to them that may become necessary. Indeed, as with other longer-existing GIs in the world, characteristics of the product and manufacturing techniques and conditions will not remain the same indefinitely but will evolve. Such evolution may entail the abandonment of some long-established cultural norms where these hinder the scale-up of production to meet increased market demand. Obviously, securing the integrity of the indication during such change is a vital consideration.

Specifications may vary in terms of how far they include bottling, packing or preparing the product for sale and consumption. Extending local control by covering every stage from cultivation of raw materials to production and presentation of the good to the final consumer is obviously highly desirable if the capacity is available to do it. For example, in Europe Parma Ham must be sliced and packaged in the designated area in order to be indicated as Parma Ham. The sale of Parma-indicated ham that was sourced appropriately but that was sliced and packaged elsewhere was the subject of a 2003 case at the European Court of Justice in which the *Consorzio del Prosciutto di Parma* (Consortium of Parma Ham) prevailed against Asda, a British supermarket chain.[22] Returning to Feni, Rangnekar suggests that in the interests both of maintaining quality and enhancing local economic returns, the specification could be modified to require the main

---

21 Rangnekar, D. 2009. *Geographical Indications and Localisation: A Case Study of Feni*. [Online]. Available at: http://papers.ssrn.com/sol3/papers.cfm?abstract_id=1564624 [accessed: 9 January 2013].

22 *Consorzio del Prosciutto di Parma & Salumificio S. Rita SpA v. Asda Stores Ltd & Hygrade Foods Ltd*, ECJ, Case C-108/01, 20 May 2003.

ingredient, cashew apples, to be sourced locally while permitting greater flexibility in terms of ingredients, distillation processes and alcoholic strength.

Reconciling the interests of all stakeholders while ensuring that fair returns go to the poorest is a far from simple matter, especially when a product is commercially successful and large corporations dominate production and distribution. In the case of Feni it is too early to say much about this.

In recent years, Tequila has seen huge growth in production, revenues and quality. And yet, the three main groups involved in the agave-tequila supply chain, the farmers, the distillers and the bottlers and distributors have not benefited equally. For one thing, under US pressure, the Mexican government decided not to ban the bulk export of Tequila mixto for bottling outside the country. This went against the interests of many farmers, distillers and government officials concerned about the loss of jobs and diminished ability to maintain quality control over the product. As Bowen put it, 'the continued exportation of tequila in bulk, despite opposition by the Mexican government, which holds the rights of the GI, reflects the power that the US-based importers and bottlers hold in the tequila industry, as well as the increasingly multinational ownership of the major tequila companies'.[23] Some Mexican Tequila firms owned by transnational corporations were also content for it to be exported in bulk.

A wider issue is the question of how far the interests of producers and consumers should take into account conservation and sustainable use of biodiversity where there may be a conflict. For example, the specification may be drawn up too narrowly so that producers are permitted to use only one crop species or variety where others could perfectly reasonably be used too. In the case of the Mexican spirit, tequila, which is the oldest non-European GI, only one agave variety may be used for 100 per cent blue agave tequilas and these are not allowed to have added sugars. In consequence, many other varieties of *Agave tequilana* are no longer being grown. In some cases, greater flexibility may be possible without jeopardizing the high reputation of the product. One approach is to allow differentiation within the indication. A good example is Scotch Whisky, a GI that accommodates differentiation into, for example, blended malts and single malts. Going back to Tequila, in addition to 100 per cent blue agave Tequilas, Tequilas mixtos need only contain 51 per cent blue agave and sugars may come from other types of plant (especially sugar cane). There are also five age categories: *joven*, *reposado*, *añejo* and *extra añejo*.[24] While such flexibility could be seen as a positive development in the sense that it may encourage the cultivation of a wider variety of agaves, allowing use of sugar from cane rather dilutes the distinctness of Tequila as a type of spirit. Indeed, this reduction in the required blue agave content suited the large producers as it lowered costs and made sugar supplies more reliable.

23 Bowen, above n 5, 239.

24 Ibid., 237.

In concluding her critical study on tequila, Bowen points to some major challenges to using GIs in developing countries for the benefit of the local stakeholders. The fundamental need is for strong legislation backed up by state involvement. The objectives of the regime should not focus narrowly on promoting the product and preventing misappropriation; rather, they should target rural development and local empowerment, and seek to prevent co-optation by transnational corporations. The more distant and powerful supply chain actors must not be allowed to shift the norms in directions that marginalize and disenfranchise the local people and dilute the indication and the reputation of the product for short-term gain. The production regulations need to be drawn up with the full participation of the local stakeholders who need to be involved in any decisions on revising production norms. This highlights a wider issue that small farmers throughout the world tend, through no fault of their own, not to be well represented as an interest group, a situation exacerbated as government generally intervene less and less in rural development matters.

The grape brandy Pisco highlights another possible complication: that two or more countries may claim the same indication on the basis that each has a similar or identical product of the same name and for which they claim to be the country of origin. This is by no means unusual. As mentioned earlier, 'Pisco' has been protected in Chile for many decades, as it has been in Peru since 1990. Chile's early designation is despite the fact that Pisco is a city and province of the neighbouring country of Peru where the drink originates. Chilean and Peruvian Pisco differ in terms of definition, grape type, alcoholic content and (obviously) designated growing area, and this does not make cooperation between the two countries at all easy. In India, the Peruvian government's application to register 'Pisco' was opposed by Chile. Accepting that the word was used in both countries, the registrar accepted the indication 'Peruvian Pisco'. In Thailand, on the other hand, Peru has succeeded in registering 'Pisco' as a GI.[25] Peru has registered 'Pisco' as an appellation of origin under the Lisbon Agreement but several state parties to the Agreement have indicated their intention of refusing the appellation on their territories, mostly because they accept Chile's right to market their drink under the same name.

In Europe, the government of Peru has registered 'Pisco' under the relevant European Union Regulation on spirits.[26] According to the technical file submitted

---

25 Trademark Office, Department of Intellectual Property, Ministry of Commerce of Thailand. 2005. *Notice of Registration of Geographical Indications Gazette (Thailand)*, 1(1). [Online]. Available at: http://www.ecap-project.org/archive/fileadmin/ecapII/pdf/en/information/thailand/gazettes/GI_Gazette_1__English.pdf [accessed: 9 January 2013].

26 Regulation (EC) No 110/2008 of the European Parliament and of the Council on the definition, description, presentation, labelling and the protection of geographical indications of spirit drinks and repealing Council Regulation (EEC) No 1576/89.

with the European GI application,[27] there are three types differentiated by type of grape and fermentation level: Pisco Puro, Pisco Acholado and Pisco Mosto Verde. As far as 'capturing' the whole supply chain, the file goes quite far, claiming 'all the stages of processing "Pisco", from growing the vine, processing the grape and bottling the final product, are carried out in the "'Pisco'-making area"'. The area is far more expansive than the city or province of Pisco, covering numerous provinces and departments.

Chile has also had successes in getting trade parties to accept the designations 'Pisco' and 'Chilean Pisco', mostly through bilateral trade agreements. While cooperation including sharing the indication would appear to be the obvious way forward, agreeing on common specifications and regulations may prove to be extremely difficult. The same may be said for basmati rice, a commercially successful product that is grown in both India and Pakistan. In both cases, achieving such cooperation is not just a technical challenge, but a political, social and cultural one too.[28] It appears that the two countries are yet to agree on a definition of what basmati rice actually is.

If it proves impossible to cooperate on the definition, what we have are two different products having the same name. The issue of homonymous products is covered in TRIPS, but only for wines. Accordingly, co-existence of both indications must be allowed. How this is to be done is left up to WTO members, but the producers concerned must be treated equitably and consumers must not be misled.[29]

A Technical Assistance Case Study: Saint Lucia

Some of the European Union's work on promoting GIs has been carried out through the TradeCom Facility, an ACP Group Programme financed by the European Development Fund that was based in Brussels. The TradeCom role was to support work in formulating and implementing trade policies and agreements, and in facilitating trade negotiations with the effective participation of APC countries. Before it closed down in spring 2012, one of TradeCom's activities was, at the request of the ACP countries, to organize workshops on GIs, commission national case studies, and produce technical reports on the subject. This part of the chapter

27 European Commission. 2011. *Main Specifications of the Technical File for 'Pisco'*. (2011/C 141/16). *Official Journal of the European Union.* [Online]. Available at: http://eur-lex.europa.eu/LexUriServ/LexUriServ.do?uri=OJ:C:2011:141:0016:0018:EN:PDF [accessed: 9 January 2013].

28 Rangnekar, D. and Kumar, S. 2010. Another look at Basmati: Genericity and the problems of a transborder geographical indication. *Journal of World Intellectual Property*, 13(2), 202-230.

29 *Agreement on Trade-related Aspects of Intellectual Property Rights*, Art 23.3.

summarizes work carried out by the author in early 2012 on the small Caribbean island state of Saint Lucia.[30]

GIs are provided under the *Geographical Indications Act* and Regulations of 2000 and 2003 respectively. Administratively, the Corporate Registrar is the head of IP for the Saint Lucia government and therefore oversees the operations of the national GI regime. The Registrar's office has plenty of experience in handling trademark applications, with about 400 granted per annum. Some 90 per cent are from foreign applications. Of the local ones, no collective or certification mark applications had been filed to date. GIs are relatively new and no applications have been made so far. Therefore there is a need for capacity building in terms of both handling initial applications and also in acquiring GIs in the first place.

In collaboration with the government and local stakeholders a number of potential GI protectable products were identified.

Products ready for immediate application

(i) Piton Water

Piton Water is a sparkling spring water product sold by the Piton Farms Mineral Water Company. Its source is the Patience Estate, which is owned by this small company. The water is collected, bottled and labelled locally. Its chemical composition is a consequence of the island state's geological features including its volcanic origin. Domestically, Piton Water competes with purified water products. Evidently, there is a lack of appreciation in Saint Lucia, including from the tourist hotels, for the specialness of Piton Water.

Looking beyond Saint Lucia, given that French culture has a high appreciation of mineral waters, the French territories of Guyane and Guadeloupe are seen as potential markets. However, according to the head of the company, marketing approval would need to come from France. This is a complicating factor, requiring documentation including on the chemical profile of the product. It would entail significant up-front cost for a small firm. The use of a GI will do nothing to alleviate this problem.

The application The owner of the company is keen to apply for a GI, and the product itself is clearly suitable for protection. The company needs to produce a specification as a component of the formal application. It needs to provide the following information:

1. The history of the product.
2. The area of production and processing.
3. Details of the product's characteristics and composition including its taste. This does not have to be so precise as to preclude the possibility of modifying the chemical ingredients.

---

30 My colleagues on the project were Bernard O'Connor and Chinwe Isiadinso.

4. Details of the production methods.
5. The geographical indication being claimed. 'Piton' refers to two iconic geological features in Saint Lucia, coastal rocky outcrops of volcanic origin. However, it should be noted that the Patience Estate is on the other side of the island from the Pitons. This does not necessarily have to be a problem. In Europe, for example, Stilton cheese is not actually made in the English village of Stilton itself, but in neighbouring districts. An acceptable alternative indication would be 'Saint Lucian mineral water' or something similar. Either way, other domestic mineral water producers on the island would be able to use the GI subject to their ability to comply with the requirements as laid out in the specification unless the fact of the source being the Patience Estate is essential to the product's uniqueness. If this is true, it would need to be justified in the specification. Either the indication would then have to be narrowed accordingly, in which case only water from the Patience Estate could use the indication. Or the Registrar might opt to provide a 'Saint Lucia' indication that could accommodate differentiation in terms of source, brand and product content just as a range of whisky products and brands made in Scotland can use the indication 'Scotch Whisky'. None of this precludes the option of the company's filing of a trade mark containing the word 'Piton' for this class of product as already exists for beer (see below).

Registration There is a clear desire on the part of the company's Managing Director to submit an application. The Corporate Registrar's office should be willing to accept a brief application that is sufficient to establish the credibility of the indication. Specifications for similar GIs in Europe can be no more than 4-5 pages in length. Clearly some consideration must be given to the wording of the indication itself. The Registrar would have the choice either of allowing the 'Piton Water' GI or establishing a more general, that is, less exclusive, indication such as 'Saint Lucian mineral water'. Moreover, as with other GIs, the government should have a graphical sign which those making and selling GI-protected products are entitled to use on the products and/or the packaging as proof of registration, as exists in Europe.

(ii) Saint Lucia Rum

Saint Lucia Distillers is a large business by Saint Lucia standards, the sole producer of rums in the country, and is a significant exporter albeit smaller than its main Caribbean rivals. The company has 27 brands (e.g. Chairman's Reserve, Bounty, Admiral Rodney) and 80 stock-keeping units and aims for the high quality end of the rum market. Such differentiation within a GI is perfectly acceptable as is evidenced by the large number of whisky products entitled to use the 'Scotch Whisky' indication. The company is regularly entered for international competitions and has won numerous medals. These are valuable for marketing purposes. There is major competition from other rum products originating in other parts of the

Caribbean, for example, Appleton, Mount Gay, Angostura and Eldorado. The company is growing rapidly and looking to expand particularly in the UK, the rest of Europe, the US and Australia. The company is doing especially well in Britain and the European Union. The latter has been cooperative in facilitating trade, bearing in mind that each country has its own labelling requirements and this incurs expense.

The application Saint Lucia Distillers is very keen to move forward and submit a domestic GIs application, and is more or less ready to apply. For Saint Lucia Distillers, GI protection is a strategic imperative given market trends. Competitors in Jamaica are moving in that direction, and this is likely to become standard practice. Although the company has plenty of experience in filing trade mark applications, it has no experience in GIs. Environment makes a huge difference in terms of taste, quality and uniqueness, and this needs to be emphasized. The water used in production is local and this is important. However, the molasses are imported, although the company has decided to start cultivating its own sugar as it had done in the past. Otherwise, the company handles all elements of the production process including bottling and so the definition should cover this too.

The company needs to produce a specification as a component of the formal application. It needs to provide the following information:

1. The history of the product.
2. The area of production and processing.
3. Details of the product's characteristics and composition.
4. Details of the production methods.
5. The geographical indication being claimed (eg 'Saint Lucia Rum').
6. In assembling the specification, the project recommended the following:
7. That the UK's Scotch Whisky Regulations 2009 be studied. These provide a great deal of relevant information on the regulation of spirits that is very specific and protects the integrity of the product, yet provides for differentiation.
8. That consideration be given to the extent to which the whole production chain from manufacture to bottling should be caught within the description. It is understood that the company has no interest at present in bulk exportation for bottling elsewhere (as is allowed under the tequila GI), but the company might wish to avoid an excessively rigid definition that would inadequately accommodate continuous differentiation and blending. It should be borne in mind that the specifications for a GI can be revised in the future subject to application and the approval of the regulators.

In the longer term, consideration may need to be given to pursuing GI protection in the main overseas markets. In this context, it is worthwhile to mention that in view of the fact that the molasses must currently be imported, and thus not all the raw materials are locally sourceable, any European application would be in the form of a Protection Geographical Indication (PGI) rather than a Protected Designation of Origin (PDO).

Registration There is a clear desire on the part of the company to submit an application. The Corporate Registrar's office should be mindful of the fact that this may well be a more complex and detailed application than some others. It should be noted that blending and mixing may be allowed and provided in the GI specification but there should be stated limits to how much adulteration is permitted. Too much will inevitably undermine the integrity and value of the GI.

(iii) Piton Beer

The Windward & Leeward Brewery, in Vieux Fort, Saint Lucia, is owned by Heineken. It supplies certain well-known brands for the whole region. Piton Beer is a very popular beer on the local market, including with visitors, though little of it is exported to date. The company licensed a brewery in Martinique to make it but the arrangement was not a success. Exporting is expensive given the weight of glass bottles, the cost of liability insurance and the marketing challenge of establishing a presence. However, licensing to distant brewers dilutes the association with the island and is not considered an attractive option. Nonetheless, there is some demand in Martinique, Guadeloupe and the UK. Given the popularity of the beer in Saint Lucia, any growth will have to come from exports. The growing popularity of Saint Lucia as a tourist destination among Europeans (especially the British) and North Americans suggests that the beer's association with the island may attract consumers to the product though this would likely entail an advertising campaign.

The application On the face of it, establishing a geographical indication appears possible due to the product's close association with the island, including its name and the local water that is used in its production. The company needs to produce a specification as a component of the formal application. It needs to provide the following information:

1. The history of the product.
2. The area of production and processing.
3. Details of the product's characteristics and composition including its taste. This does not have to be precise. The fact of the water being locally sourced should help to justify the uniqueness of the product.
4. Details of the production methods. Again, a precise description is unnecessary.
5. The geographical indication being claimed. As with the spring water, given that 'Piton' refers to well-known geographical features in Saint Lucia, 'Piton Beer', which is already part of a registered trade mark, would be acceptable and might be highly attractive. However, it should be noted that

the location of the brewery is some distance from the Pitons as depicted on the label.

Registration There is a clear interest in submitting an application. The Corporate Registrar's office should be willing to accept a brief application which is sufficient to establish the credibility of the indication. Specifications for similar GIs in Europe can be no more than 4-5 pages in length. Clearly some consideration must be given to the wording of the indication itself. The Registrar would have the choice either of allowing the 'Piton Water' GI or establishing a more general, that is, less exclusive, indication such as 'Saint Lucian beer'.

(iv) Handicrafts

The handicrafts sector generates employment for hundreds of Saint Lucians and several million US dollars in income. Probably the best-organized producers are those who have formed the Choiseul Association of Craft and Heritage Tourism (CATCH). Choiseul is a district and village in the South-east of Saint Lucia and is known for its arts and crafts. There are numerous distinctive handicraft products made using raw materials sourced exclusively in Saint Lucia. These include coal pots, sisal handbags and mats, and tatanye brooms. CATCH has been very proactive having already prepared a mark that it aims to register in Saint Lucia.

Application CATCH is keen to pursue GI protection and, as mentioned, has already produced a mark it intends to register as a trademark. Although protecting handicrafts by a GI is not possible in Europe, it is possible in Saint Lucia as it is in some other countries including India. However, some consideration would need to be given as to whether the indication should be for 'Choiseul' or 'Saint Lucia'. Of course a Choiseul indication would not preclude also the use of such a phrase as 'Genuine Saint Lucia Hand Crafted Product' on the label.

An association seeking to register a GI would need to produce a specification as a component of the formal application. It needs to provide the following information:

1. The history of handicrafts in Saint Lucia and/or the locality as appropriate including each individual type of product.
2. The area of production and processing, whether Choiseul or the country as a whole.
3. Details of the type of handicraft products for which the GI would be used including the materials used and where they are sourced.
4. Details of the production methods. These should be primarily if not entirely hand-made.

Registration There is a clear interest in submitting an application from CATCH. There is little to lose from doing so and much to gain. Many tourists wish to be assured that a handicraft is not made in China or another distant country but is genuinely local. The Corporate Registrar's office should be willing to accept a brief application which is sufficient to establish the credibility of the indication. Some consideration must be given to the wording of the indication itself. In the future the government might consider adopting a labelling scheme authenticating products that would comprise words and/or a symbol as in some other countries.

Products not ready for immediate application but worthy of consideration

(i) Cocoa and cocoa products

Saint Lucia needs to reduce its dependence on bananas by diversifying its agricultural production and identifying new high-value low-volume products. There are some five cocoa plantations on the island. Cocoa is also produced on small- and medium-sized farms. There are a number of cocoa-based products: raw (wet) cocoa; cocoa fat; cocoa paste; cocoa powder; and cocoa sticks and balls; and of course chocolate. Consequently, cocoa and cocoa products could be protected in various forms as product in themselves or else as an ingredient of products such as chocolate that may be manufactured elsewhere. It should be noted that cocoa from Saint Lucia is not markedly different from cocoa produced on other Caribbean islands. Moreover, it may be blended with cocoa from elsewhere in the region or in the world. In this context it is noteworthy that ingredients and not just finished products can be GI-protected as long as the GI is not misused in a way implying that it covers the whole product.

However, the fermentation process employed may result in some minor differences that a GI could highlight. Admittedly there is nothing particularly technical about the method. Otherwise, the use of trademarks might be preferable.

Before going further, it is important to consider whether and how Saint Lucian cocoa and cocoa products can realistically be differentiated from cocoa from the neighbouring islands, and whether a strategy of differentiation would be beneficial anyway. One possibility might be to discuss the possibility of a regional GI that would allow member of the Caribbean Fine Cocoa Forum to use it in those states where it may be registered. Moreover, it is not in any case a foregone conclusion that producers in Saint Lucia stand to gain from a GI given that they are already able to sell their whole production at good prices. One quality chocolate manufacturer with interests in Saint Lucia, Hotel Chocolat, indicates the Saint Lucia source of the cocoa used in some of its chocolate products and this serves to build up goodwill. The company intends to start manufacturing chocolate on the island.

(ii) Other products

Several other products were mentioned as worthy of investigation as possible Saint Lucian GIs:

- lansan-based essential oils and incense
- coconut oils and drinks
- herbal formulations

- cassava
- cinnamon
- citrus fruits
- capsicum
- sea moss

Wider production of some of these may raise concerns about sustainability. For example, the lansan tree, which is endemic to the region, has become scarce in several of the islands due to overexploitation of the resin. It is currently on the IUCN Red List of Threatened Species. Any effort to promote commercialization of such products including medicinal plants should give attention to: (i) conservation, (ii) the optimization of resource harvesting and cultivation practices so that wider utilization is not at the expense of the environment, and (iii) equitable benefit sharing with local communities in conformity with the *Convention on Biological Diversity*.

Recommendations There are some obvious differences here. Some of these products are very local in terms of their content and the degree to which their production involves uniquely Saint Lucian natural and human resources. Others are far less local. Indeed, some producers are hardly local at all. The alcoholic beverages sector is one characterized by international businesses with experience in marketing in various countries and holding and managing substantial intellectual property portfolios, especially trademarks. These firms can look after themselves with or without GIs being available in the local market – though having one may certainly be helpful given the need to have domestic protection for legal means to prevent misuse in other jurisdictions. The jury is definitely out as to whether small producers can truly help small firms and community growers and producers. This is especially so when the local market is small with limited scope for expansion as is the case in Saint Lucia and quite a lot of other developing countries, excellent as some of these products undoubtedly are. Exporting for these businesses is unlikely to be a straightforward matter. Different countries offer a range of regulatory challenges to trade such as those regarding content, production methods, customs procedures, packaging and tariffs. All of these can make exporting complex, risky and expensive. Regulations are especially strict in the case of products for human consumption, such as foods and beverages. The difficulties in exporting mineral water to the Francophone Caribbean islands, one of which is visible on a clear day from Saint Lucia, are sadly typical.

The project closed with the following list of recommendations to the government and to producers.

Government

- Discussions with stakeholders suggest that some producers are confused about the application process and this may be deterring the submission of applications. There is a need to encourage applications. This points to the importance of demystifying the GIs application process by clarifying the procedures. More outreach is necessary to further enhance awareness.

- Accordingly, the Corporate Registrar's office should seek to engage with producer associations involved with any of the above mentioned products in order to enhance awareness and to encourage the submission of applications for registration as appropriate.
- The government and donors should support the Registrar's office by providing sufficient resources so that applications can be processed in a speedy fashion, and to enhance the necessary outreach work so that the Registrar's office can engage more effectively with the producers.
- Given the public nature of GIs, consideration should be given to removing all fees and charges for the registration of GIs. Specific financing could be requested within the context of the Cooperation provisions of the *Economic Partnership Agreement with the European Union* (EPA).
- A GI logo needs to be designed urgently. It should contain the terms 'Saint Lucia' and 'Geographical Indication'. Ideally, its design should be evocative of the country.
- The *Geographical Indications Act* of 2000 will have to be amended by 1 January 2021. Assistance in the redrafting of the Act could be provided under the IP aid and cooperation provisions of the EPA.

Producers

- In general, applying for a GI is neither an expensive nor a difficult process. Producers can request the guidance of the Registrar's office while preparing a formal application.
- GIs are not a panacea. Attention must always be given to other means of maintaining trust and high reputation, such as quality control, appropriate pricing, attractive packaging and marketing.
- Trademarks should not be discounted as an alternative, or in addition to, GIs.
- In some cases it may be preferable to apply for a GI through a producer association rather than individually. This will require a cooperative relationship among all producers sharing the rights to use the indication including a collective commitment to continuing to produce according to the specification.

Conclusion

In a recent report for the International Trade Centre, Giovannucci et al. point to the pros and cons of GIs, and offer four conditions for success which are generally applicable also to collective and certification trademarks.[31] One the one hand, they 'can offer a comprehensive framework for rural development since they can positively encompass issues of economic competitiveness, stakeholder equity, environmental stewardship, and socio-cultural value'. On the other, especially if the GI is poorly designed or inadequately governed, 'badly managed GIs can be dominated by limited political interests or just a few enterprises. In some cases, GIs can exclude the poorest producers or even stimulate inappropriate outcomes such as the dissolution of traditional practices or the destruction of biodiversity.' As for the conditions for success, the authors of the above report offer the following:

1. *Strong organizational and institutional structures* to maintain, market, and monitor the GI. The core processes of: (i) identifying and fairly demarcating a GI (ii) organizing existing practices and standards and (iii) establishing a plan to protect and market the GI all require building local institutions and management structures with a long-term commitment to participatory methods of cooperation.
2. *Equitable participation* among the producers and enterprises in a GI region. Equitable is here defined as the participating residents of a GI region sharing reasonably in not only costs and benefits but also in the control and decisions regarding their public assets.
3. *Strong market partners* committed to promote and commercialize over the long term. Many of the GI market successes are the result of mutually beneficial business relations via which consistent market positioning and effective commercialization have led to a long-term market presence.
4. *Effective legal protection* including a strong domestic GI system. Carefully chosen protection options will permit effective monitoring and enforcement in relevant markets to reduce the likelihood of fraud that can compromise not only the GI's reputation but also its legal validity.

This chapter has sought to elucidate some of the trade and development implications of GIs in the developing country context. This has been a challenge, first, because giving legal effect to the term 'geographical indication' is far from straightforward given their legally rather fuzzy nature. In this respect they are quite different from patents, copyright and trademarks. Second, GIs have not been used much in developing countries. Indeed, most of the world's registered GIs are European. The confusion and uncertainty about GIs is partly reflected in the lack of international consensus as to how they should be regulated and whether indications should be treated differently under the law according to what type of product is to be protected.

---

31 Giovannucci, D., Josling, T., Kerr, W., O'Connor, B. and Yeung, M.T. 2009. *Guide to Geographical Indications: Linking Products and their Origins.* [Online]. Available at: http://www.intracen.org/publications/Free-publications/Geographical_Indications.pdf [accessed: 9 January 2013].

Nonetheless, there is sufficient experience now to demonstrate that GIs have the potential to promote certain goods that can generate income for local communities and small-scale producers in developing countries. But GI regimes can be complicated. The market alone cannot regulate the system and ensure it works for the benefit of all stakeholders especially the poorer ones. For one thing, GIs are not private rights (unless they are in the form of trademarks), and can thus be seen as public assets. Moreover, a certain amount of state involvement is necessary for a registration system to operate and in case producer associations lack the capacity to police misuse of the indication in foreign markets. These features may lead some communities to opt for collective trademark protection. Trademarks are private property, and their registration is more likely to be inexpensive, straightforward and un-bureaucratic once the mark itself has been created by the community association that will make the application. Indeed, this is why an association representing indigenous communities inhabiting the Potato Park in highland Peru has recently chosen to eschew seeking GI protection in favour of developing its own mark which it is currently using, so far with some success.

The production specifications are of vital importance if consideration is to be given to the needs of local producers, the importance of capturing as much of the production chain to keep as much of the value in the region as possible (taking into account possible capacity deficiencies), and the vital need to maintain the integrity of the indication. However, a certain amount of flexibility should be allowed. Finally, honest and competent government, political stability and a conducive business environment are preconditions for any regulatory system to operate properly. What is true for any such system is true for GIs.

Chapter 10
Patent-Busting: The Public Patent Foundation, Gene Patents and the Seed Wars

Matthew Rimmer*

Introduction

In the United States, there has been much concern about the operation of the patent system – with concerns raised about the quality of the patents granted by the United States Patent and Trademark Office (USPTO), and the enforcement of patents in the courts.[1] There has been a particular concern about the rise of 'patent trolls' – strategic non-practising entities which use patents to threaten others with the Hobbesian choice of paying licensing fees or facing patent infringement proceedings.

In reaction to the perceived malaise of the patent system, there has been an emergence of 'patent-busters', which seek to challenge the validity of patents of dubious novelty, inventiveness and utility. The Public Patent Foundation (PUBPAT for short) is a not-for-profit legal services organization based at the Benjamin N. Cardozo School of Law that 'represents the public's interests against the harms caused by the patent system, particularly the harms caused by undeserved patents and unsound patent policy'.[2] The organization has observed: 'Undeserved patents injure the public because they can be used by private actors to preclude activity that would otherwise be permissible, if not desirable.'[3] The Foundation has sought to represent the larger public interest in patent law through legal representation, public education and advocacy.

The director of the Public Patent Foundation is Daniel Ravicher. He has been labelled a modern day 'Robin Hood' by *Science* magazine.[4] His interest in patent law was piqued by classes at the University of Virginia by Randall Rader, a chief

* The author thanks Sabrina Hassan, senior counsel for the Public Patent Foundation, for her discussions on this topic.

1 See, for instance, Burk, D. and Lemley, M. 2009. *The Patent Crisis and How the Courts Can Solve It*. Chicago and London: University of Chicago Press.

2 The Public Patent Foundation. 2012. *About PUBPAT*. [Online]. Available at: http://www.pubpat.org/About.htm [accessed: 1 February 2013].

3 Ibid.

4 Kintisch, E. 2005. Dan Ravicher profile: A 'Robin Hood' declares war on lucrative US patents. *Science*, 309(5744), 2166.

judge of the Federal Circuit Court of Appeals. In an interview with Groklaw, Ravicher explains the reasons for establishing the foundation:

> The patent system is being abused by private actors to the detriment of the mostly unaware public. Our health, our freedom, and our economic prosperity are all under assault from bogus rights meted out to the few with the power and expertise to game a system originally established hundreds of years ago to promote progress within society as a whole. The government, through primarily a captured patent office utterly failing to achieve its mission and skewed policies implement into patent law by Congress and the courts, is not just failing to defend the public interest from abuse of the patent system, but is complicit in and supportive of such efforts. In information technology industries, abuse of the patent system means illicit restraint of civil liberties and unjustified disproportionate burdening of small businesses. In life science industries, abuse of the patent system has even more devastating results, including the exacerbation of pain and suffering by those who cannot afford medical technologies covered by undeserved patents. This situation is abhorrent and the Public Patent Foundation is beginning a campaign against such abuses.[5]

The influential magazine *Managing Intellectual Property* rated Ravicher one of the 50 Most Influential People in Intellectual Property in 2012.[6]

The Public Patent Foundation's Daniel Ravicher envisaged:

> PubPat's four core activities are (1) challenging patents that threaten the public's health, freedom, or other interests, (2) helping small businesses defend themselves from patents being asserted against them, (3) establishing patent commons within markets crippled by patent thickets, and (4) educating the public regarding these issues and advocating for reform of the patent system.[7]

Ravicher explains that the most difficult part of his job is encouraging the legal profession to have an open mind about his work:

> People like to say: Oh, that guy he hates all patents. It could not be further from the truth. That's like saying the guy from The Innocence Project hates criminal law… I felt nobody was representing the public interest. There was

---

5 Jones, P. 2003. Interview with Public Patent Foundation's Dan Ravicher. *Groklaw*. [Online]. Available at: http://www.pubpat.org/LWN_Interview.htm [accessed: 1 February 2013].

6 Managing Intellectual Property. 2012. *Dan Ravicher, Public Patent Foundation: He Hates Bad Patents*. [Online]. Available at: http://www.managingip.com/Article/3060577/Dan-Ravicher-Public-Patent-Foundation-He-hates-bad-patents.html?edit=true [accessed: 1 February 2013].

7 Jones, above n 5.

> nobody standing up saying that sometimes the Patent Office goes too far or patent holders go too far.[8]

The Public Patent Foundation has earned a certain amount of ire from the patent profession and related industries. Susan Hansen notes: 'For a guy who gave up a high-paying corporate law job to try to improve the patent system, Daniel Ravicher doesn't always get a lot of thanks'.[9]

The Public Patent Foundation carefully selects the patents that it is challenging. According to *Science*, Ravicher 'supervises a handful of volunteer scientists, occasional grad students, and legal interns as they search for potential flaws in big-name patents'.[10] He particularly targets the patents that he believes 'are causing the most harm'.[11] In a profile, Crispin Littlehales notes that Ravicher uses 'sources willing to give him tips – usually disgruntled scientists – to help identify his challenges'.[12] Moreover, he observes that Ravicher 'relies on the wisdom of a like-minded board of directors and several advisors who believe in what PUBPAT does but do not want to be publicly affiliated with it for political and other reasons'.[13] The board of directors for the Public Patent Foundation currently include Professor Margo Bagley, Professor Michael Herz, Brian Kahin, Professor Eben Moglen, the General Counsel for the Free Software Foundation and Associate Professor Joshua Sarnoff.

The group has had some success in filing requests for re-examination of patents at the USPTO. The Public Patent Foundation has been active in the area of information technology patents. In 2004, the Foundation presented prior art which invalidated Microsoft Corporation's patent on the FAT File System, which was promoted as 'the ubiquitous format used for interchange of media between computers, and, since the advent of inexpensive, removable flash memory, also between digital devices'.[14] In 2008, the Public Patent Foundation sought a review of a patent on photomosaics, which are pictures made through the use of many small pictures.[15] There had been concern that the patentee, Robert Silvers, had been threatening to assert the patent against individuals, small businesses and even open source developers. In 2010, the patent examiner issued a final rejection of the majority of the patent's claims.

---

8 Managing Intellectual Property, above n 6.

9 Hansen, S. 2007. Power to the people. *IP Law And Business*, April, 36-42, 36.

10 Kintisch, above n 4.

11 Ibid.

12 Littlehales, C. 2008. Profile: Dan Ravicher. *Nature Biotechnology*, 26(4), 369.

13 Ibid.

14 The Public Patent Foundation. 2012. *Microsoft FAT Patent.* [Online]. Available at: http://www.pubpat.org/microsoftfat.htm [accessed: 1 February 2013].

15 The Public Patent Foundation. 2012. *Silvers Photomosaic Patent.* [Online]. Available at: http://www.pubpat.org/silvers.htm [accessed: 1 February 2013].

In 2004, the Public Patent Foundation filed a request for re-examination in relation to a 'co-transformation' patent – this is a process for inserting foreign DNA into a host cell to produce certain produce certain proteins, which are the basis for a wide range of pharmaceutical products.[16] The underlying technology was relevant for Epogen for anaemia, Activase for heart attacks strokes, Avonex for multiple sclerosis and recombinant for Haemophilia. The USPTO and Columbia University waived any right to assert the patent in December 2004. The Public Patent Foundation has also had a keen interest in patent law and biotechnology. In 2006, the Foundation filed formal requests with the Patent Office on behalf of the Foundation for Taxpayer and Consumer Rights to revoke three patents owned by the Wisconsin Alumni Research Foundation in respect of stem research. The Patent office rejected all claims of each of the patents on March 30, 2007.[17]

The Public Patent Foundation has also been active in the field of patent law, pharmaceutical drugs and access to essential medicines.[18] In 2004, the Public Patent Foundation submitted prior art, challenging the validity of Pfizer's patent on Lipitor, a treatment for lowering cholesterol.[19] As a result, the USPTO rejected the patent in June 2005, and Pfizer conceded its original broad claims. In 2007 and 2008, the Public Patent Foundation challenged key HIV/AIDS pharmaceutical drug patents held by Gilead Sciences.[20] The USPTO rejected all of the claims of each of the patents. In 2010, the Public Patent Foundation challenged eight patents held by Abbott Laboratories for the key HIV/AIDS drug ritonavir (which has been marketed under the trade name, Norvir).[21] This challenge is under consideration by the USPTO. In the 2011 matter of *Public Patent Foundation Inc. v. GlaxoSmithKline Consumer Healthcare LP*, the Public Patent Foundation filed a complaint against GlaxoSmithKline Consumer Healthcare LP, alleging that the

---

16 The Public Patent Foundation. 2012. *Co-Transformation Patent.* [Online]. Available at: http://www.pubpat.org/columbiacotransformation.htm [accessed: 1 February 2013].

17 The Public Patent Foundation. 2012. *WARF Stem Cell Patents.* [Online]. Available at: http://www.pubpat.org/warfstemcell.htm [accessed: 1 February 2013].

18 For an overview of the debate over patent law, and access to essential medicines, see Abbott, F. and Dukes, G. 2009. *Global Pharmaceutical Policy: Ensuring Medicines for Tomorrow's World*, Cheltenham and Northampton, MA: Edward Elgar; Pogge, T., Rimmer, M. and Rubenstein, K. (editors) 2010. *Incentives for Global Public Health: Patent Law and Access to Medicines*. Cambridge: Cambridge University Press; Ho, C. 2011. *Access to Medicine in the Global Economy: International Agreements on Patents and Related Rights*, Oxford: Oxford University Press; Matthews, D. 2011. *Intellectual Property, Human Rights and Development*, Cheltenham and Northampton, MA: Edward Elgar.

19 The Public Patent Foundation. 2012. *Pfizer Lipitor Patent.* [Online]. Available at: http://www.pubpat.org/pfizerlipitor.htm [accessed: 1 February 2013].

20 The Public Patent Foundation. 2012. *Gilead HIV/AIDS Drug.* [Online]. Available at: http://www.pubpat.org/gileadhivaidsdrug.htm [accessed: 1 February 2013].

21 The Public Patent Foundation. 2012. *Ritonavir HIV/AIDS Drug Patents.* [Online]. Available at: http://www.pubpat.org/ritonavir.htm [accessed: 1 February 2013].

company had violated the False Marking Statute, since January 2005, 'by marking certain of its products with the patent numbers of expired patents'.[22] The district court judge held that the licensee had no knowledge that patents marking products had expired, and, even assuming the licensee's knowledge, the products were not marked with expired patents to deceive the public.

This chapter considers the Public Patent Foundation as a novel institution in the patent framework. It contends that such a model can play a productive role in challenging the validity of high-profile patents; working as amicus curiae in significant court cases; and also promoting patent law reform. However, there are limits to the 'patent-busting' of the Foundation. The not-for-profit legal services organization has only had the time and resources to challenge a number of noteworthy patents. Other jurisdictions – such as Australia – lack such public-spirited 'patent-busting' entities. This chapter considers a number of key disputes involving the Public Patent Foundation. Part I examines the role of the Public Patent Foundation in the landmark dispute in the United States over Myriad Genetics' patents in respect of breast cancer and ovarian cancer.[23] Part II considers the role of the Public Patent Foundation in litigation between organic farmers and Monsanto in the United States.[24] Part III examines the role of the Public Patent Foundation in larger debates about patent law reform in the United States – particularly looking at the *Leahy-Smith America Invents Act* 2011 (US).[25] The conclusion contends that the patent-busting model of the Public Patent Foundation should be emulated in respect of other technological fields, and other jurisdictions – such as Australia. The initiative could also be productively applied to other forms of intellectual property – such as trade mark law, designs law, plant breeders' rights and access to genetic resources.

I. Gene Patents

There has been longstanding controversy over Myriad Genetics and its patents in respect of genetic testing for breast cancer and ovarian cancer.[26] The Public Patent

22 *Public Patent Foundation Inc.* v. *GlaxoSmithKline Consumer Healthcare* LP 801 F. Supp. 2d 249 (2011).

23 *Association for Molecular Pathology* v. *USPTO and Myriad Genetics Inc.* 702 F. Supp. 2d 181 (S.D.N.Y., 2010); *Association for Molecular Pathology* v. *USPTO and Myriad Genetics Inc.* 653 F. 3d 1329 (C.A. Fed. (N.Y.) 2011); *Association for Molecular Pathology* v. *USPTO and Myriad Genetics Inc.* 132 S. Ct. 1794. (2012); *Association for Molecular Pathology* v. *USPTO and Myriad Genetics Inc.* F. 3d, 2012 WL 3518509.

24 *Organic Seed Growers and Trade Association* v. *Monsanto Company and Monsanto Technology LLC* 851 F. Supp. 2d 544 (2012).

25 *Leahy-Smith America Invents Act*, 2011 (US) Pub. L. No. 112-29.

26 See, for instance, Rimmer, M. 2003. Myriad Genetics: Patent law and genetic testing. *European Intellectual Property Review*, 25(1), 20-33; Rimmer, M. 2008. *Intellectual Property and Biotechnology: Biological Inventions*, Cheltenham and

Foundation collaborated with the American Civil Liberties Union in mounting a challenge in the United States courts against Myriad Genetics' patents in respect of BRCA1 and BRCA2, which are related to breast cancer and ovarian cancer. In May 2009, the two groups filed a lawsuit 'charging that patents on two human genes associated with breast and ovarian cancer stifle research that could lead to cures and limit women's options regarding their medical care'.[27] Discussing the legal action, Daniel Ravicher of the Public Patent Foundation emphasized: 'Patenting human genes is counter to common sense, patent law and the Constitution. Genes are identified, not invented, and patenting genetic sequences is like patenting blood, air or $e=mc^2$.'

The legal action was filed on behalf of women patients, cancer survivors, scientific researchers, health groups and scientific associations, representing pathologists, scientists and researchers. The plaintiffs provided powerful testimony about the impact of the patents upon patient health, scientific research and the health-care system.[28] The legal action was filed against the USPTO, as well as Myriad Genetics and the University of Utah Research Foundation. The epic litigation has progressed from a District Court of the United States; to the Court of Appeals for the Federal Circuit; and has been appealed to the Supreme Court of the United States.

The United States District Court for the Southern District of New York

At first instance, Sweet J of the United States District Court for the Southern District of New York was sympathetic to the arguments of the Public Patent Foundation and the American Civil Liberties Union.[29] The judge observed of the conflict:

Northampton, MA: Edward Elgar; Stroud, J. 2012. A thousand tiny pieces: The Federal Circuit's fractured Myriad ruling, lessons to be learned, and the way forward. *IP Theory*, 2(2), 71-82; Rimmer, M. and McLennan, A. (editors). 2012. *Intellectual Property and Emerging Technologies: The New Biology*, Cheltenham and Northampton, MA: Edward Elgar; Huys, I., Matthijs, G. and Van Overwalle, G. 2012. The fate and future of patents on human genes and genetic diagnostic methods. *Nature Reviews Genetics*, 13(6), 441-448.

27 The American Civil Liberties and the Public Patent Foundation. 2009. *ACLU and PUBPAT Challenge Patents on Breast Cancer Genes: Gene Patents Stifle Patient Access To Medical Care And Critical Research.* [Online]. Available at: http://www.pubpat.org/brcafiled.htm [accessed: 1 February 2013].

28 American Civil Liberties Union. 2012. *ACLU Challenges Patents on Breast Cancer Genes: BRCA.* [Online]. Available at: http://www.aclu.org/free-speech-womens-rights/aclu-challenges-patents-breast-cancer-genes-0 [accessed: 1 February 2013].

29 *Association for Molecular Pathology* v. *USPTO and Myriad Genetics Inc.* 702 F. Supp. 2d 181 (S.D.N.Y., 2010).

> Plaintiffs' challenge to the validity of these claims, and the arguments presented by the parties and amici, have presented a unique and challenging question: Are isolated human genes and the comparison of their sequences patentable? Two complicated areas of science and law are involved: molecular biology and patent law. The task is to seek the governing principles in each and to determine the essential elements of the claimed biological compositions and processes and their relationship to the laws of nature. The resolution of the issues presented to this Court deeply concerns breast cancer patients, medical professionals, researchers, caregivers, advocacy groups, existing gene patent holders and their investors, and those seeking to advance public health.[30]

The judge commented: 'The resolution of these motions is based upon long recognized principles of molecular biology and genetics: DNA represents the physical embodiment of biological information, distinct in its essential characteristics from any other chemical found in nature.'[31] Sweet J 'concluded that DNA's existence in an "isolated" form alters neither this fundamental quality of DNA as it exists in the body nor the information it encodes'.[32] The judge ruled that 'the patents at issue directed to "isolated DNA" containing sequences found in nature are unsustainable as a matter of law and are deemed unpatentable subject matter under 35 U.S.C. § 101'.[33] The judge also ruled that, 'because the claimed comparisons of DNA sequences are abstract mental processes, they also constitute unpatentable subject matter under § 101'.[34]

In the course of the judgment in his discussion of law and genetics, Sweet J refers back to debates over agriculture as a touchstone. The judge provides a portrait of Mendelian genetics:

> The field of genetics – the science of heredity and variation in living organisms – and the concept of units of heredity that could be transmitted from one generation to another originated in the 19th century from experiments with pea plants conducted by Gregor Mendel. Mendel showed that certain traits are passed on from parent to offspring as discrete entities and do not appear blended in the offspring. He hypothesized that it was the plant's genotype, or assortment of hereditary factors, that determined the plant's phenotype, or appearance. In 1909, this unit of inheritance was termed a 'gene'. Yet the gene remained an abstract concept until 1915, when it was shown that genes corresponded to physical spans of chromosomal material.[35]

30 Ibid., 184.
31 Ibid., 185.
32 Ibid.
33 Ibid.
34 Ibid.
35 Ibid., 192.

The judge observed:

> While the Supreme Court in *J.E.M. Ag Supply*[36] noted the USPTO's practice of issuing patents on sexually reproducing plants in concluding that such plants represented patentable subject matter under § 101, that passing observation was neither dispositive nor central to the Court's holding and does not establish a rule of judicial deference to the USPTO's practices.[37]

Lauding the decision by Sweet J, Daniel Ravicher wrote about the potential implications of the ruling on gene patents in *The Huffington Post*.[38] He maintained: 'To say that the Honorable Robert Sweet of the Southern District of New York saved the lives of millions of Americans on Monday with the stroke of a pen is not a hyperbole.'[39] Ravicher expressed the hope that the decision would have a positive impact upon the burgeoning field of personalized medicine:

> Our victory this week will enable a transformative change in medicine to the benefit of patients, society and, yes, even industry. In the not too distant future, when you go to the doctor, she will screen your entire DNA to determine which medicines would work best for you and which would be harmful. This will become as common as taking your temperature and running other blood work. The now realizable prospect of leaving behind the one-size-fits-all world of medical treatment for custom-tailored diagnosis and treatment should make all Americans breath a deep sigh of relief. And it would not be possible if gene patents were upheld, because then the cost of a full genetic screening would be prohibitively high due to the hold up incentives placed on patent holders, who are driven to maximize profit, not public good.[40]

In his 2012 book on the future of intellectual property and health-care, *Identity and Invention*, Shubha Ghosh explores the culture and ethics of patent law and personalized medicine.[41] He notes in his conclusion that 'Medical information is as much a part of the information age as the Internet, software, and social networking.'[42] There is certainly a larger interest in the impact of the litigation

---

36 *J.E.M. Ag Supply* v. *Pioneer Hi-Bred International Inc* 534 US 124 (2001).

37 *Association for Molecular Pathology* v. *USPTO and Myriad Genetics Inc*. 702 F. Supp. 2d 181 at 221 (S.D.N.Y., 2010).

38 Ravicher, D. 2010. Gene Patent Decision Emancipates Personalized Medicine. *The Huffington Post*. [Online]. Available at: http://www.huffingtonpost.com/daniel-b-ravicher/gene-patent-decision-eman_b_521324.html [accessed: 1 February 2013].

39 Ibid.

40 Ibid.

41 Ghosh, S. 2012. *Identity and Invention: The Culture and Ethics of Personalized Medicine Patenting*, Cambridge: Cambridge University Press.

42 Ibid.

upon intellectual property and a variety of fields of biotechnology – cutting across the agriculture, health-care, medicine, the environment, and climate mitigation and adaptation.

The United States Court of Appeals for the Federal Circuit

The United States Court of Appeals for the Federal Circuit reviewed the decision of Sweet J.[43] Lourie J for the majority summarized the complex ruling of the court:

> On the threshold issue of jurisdiction, we affirm the district court's decision to exercise declaratory judgment jurisdiction because we conclude that at least one plaintiff, Dr. Harry Ostrer, has standing to challenge the validity of Myriad's patents. On the merits, we reverse the district court's decision that Myriad's composition claims to 'isolated' DNA molecules cover patent-ineligible products of nature under § 101 since the molecules as claimed do not exist in nature. We also reverse the district court's decision that Myriad's method claim to screening potential cancer therapeutics via changes in cell growth rates is directed to a patent-ineligible scientific principle. We, however, affirm the court's decision that Myriad's method claims directed to 'comparing' or 'analyzing' DNA sequences are patent ineligible; such claims include no transformative steps and cover only patent ineligible abstract, mental steps.[44]

The judge emphasized: 'If the law is to be changed, and DNA inventions excluded from the broad scope of § 101 contrary to the settled expectation of the inventing community, the decision must come not from the courts, but from Congress.'[45]

In a separate judgment, Moore J was reluctant to disrupt and unsettle established practice in respect of patent law and biotechnology:

> This case typifies an observation by the late Chief Judge Markey, our first Chief Judge, that '[o]nly God works from nothing. Men must work with old elements.' Human DNA is, for better or worse, one of the old elements bequeathed to men to use in their work. The patents in this case revealed a new molecular understanding about ourselves; 'the inventions most benefiting mankind are those that "push back the frontiers of chemistry, physics, and the like".' We cannot, after decades of patents and judicial precedent, now call human DNA fruit from the poisonous tree, and punish those inquisitive enough to investigate, isolate, and patent it. 'Our task … is the narrow one of determining what Congress meant by the words it used in the statute; once that is done our powers are exhausted.' This inquiry does not have moral, ethical, or theological components. *Cf. id.* ('[W]e

43 *Association for Molecular Pathology* v. *USPTO and Myriad Genetics Inc.* 653 F. 3d 1329 (C.A. Fed. (N.Y.) 2011).

44 Ibid., 1333-1334.

45 Ibid., 1357.

are without competence to entertain' arguments about 'the grave risks' generated by genetic research.). The patents in this case might well deserve to be excluded from the patent system, but that is a debate for Congress to resolve.[46]

The judge declined 'to extend the "laws of nature" exception to include isolated DNA sequences'.[47]

In dissent, Bryson J maintained that Myriad Genetics should be able to obtain patent protection in relation to genes:

> In its simplest form, the question in this case is whether an individual can obtain patent rights to a human gene. From a common-sense point of view, most observers would answer, 'Of course not. Patents are for inventions. A human gene is not an invention.' The essence of Myriad's argument in this case is to say that it has not patented a human gene, but something quite different – an *isolated* human gene, which differs from a native gene because the process of extracting it results in changes in its molecular structure (although not in its genetic code). We are therefore required to decide whether the process of isolating genetic material from a human DNA molecule makes the isolated genetic material a patentable invention. The court concludes that it does; I conclude that it does not.[48]

The judge emphasized that, while 'it may be very difficult to extract the newly found mineral or to find, extract, and propagate the newly discovered plant', 'that does not make those naturally occurring items the products of invention'.[49] Bryson J observed: 'The same is true for human genes.'[50] The judge emphasized: 'In each case, merely isolating the products of nature by extracting them from their natural location and making those alterations attendant to their extraction does not give the extractor the right to patent the products themselves.'[51]

Considering the outcome, Daniel Ravicher stressed for the Public Patent Foundation: 'As all the judges today agreed that our clients had standing to bring this case and that analyzing gene sequences for alterations is not patentable.'[52] He expressed his preference for the dissent of Bryson J: 'The judges disagreed with each other on whether pieces of the human genome are patentable, and we agree

---

46 Ibid., 1372-1373.
47 Ibid., 1373.
48 Ibid.
49 Ibid., 1375.
50 Ibid.
51 Ibid.
52 The Public Patent Foundation. 2011. *Appeals Court Unanimously Invalidates Patents on Human Gene Analysis, Divides of Whether Genes Themselves May Be Patented.* [Online]. Available at: http://www.pubpat.org/cafcpaneldecision.htm [accessed: 1 February 2013].

with Judge Bryson who explained they are not because no one "invents" genes.'[53] Ravicher reiterated his case: 'Inventions are things like new genetic tools or drugs, all of which can be patented because they are not genes themselves.'[54]

The Supreme Court of the United States and the Prometheus Decision

In December 2011, the American Civil Liberties Union and the Public Patent Foundation requested that the Supreme Court of the United States consider an appeal against the ruling of the Court of the Appeals for the Federal Circuit.[55]

In the interim, the Supreme Court of the United States ruled upon the nature of patentable subject matter in the *Prometheus* matter.

In the 2012 case of *Mayo Collaborative Services* v. *Prometheus Laboratories Inc.*, the Supreme Court of the United States rejected a patent application in respect of medical information on the grounds that it claimed the underlying laws of nature.[56] This decision marked an important shift from an earlier consideration of the matter.[57] In his lead judgment for the Supreme Court of the United States, Breyer J re-articulates the law with respect to patentable subject matter: 'If a law of nature is not patentable, then neither is a process reciting a law of nature, unless that process has additional features that provide practical assurance that the process is more than a drafting effort designed to monopolize the law of nature itself.'[58] Focusing on 'laws of nature', Breyer J comments: 'The Court has repeatedly emphasized ... a concern that patent law not inhibit further discovery by improperly tying up the future use of laws of nature'.[59] His Honour noted: 'These statements reflect the fact that, even though rewarding with patents those who discover new laws of nature and the like might well encourage their discovery, those laws and principles, considered generally, are "the basic tools of scientific and technological work".'[60] Breyer J was concerned about the danger 'that the

53 Ibid.

54 Ibid.

55 The Public Patent Foundation. 2011. *PUBPAT and ACLU Ask Supreme Court to Rule that Patents on Breast Cancer Genes are Invalid.* [Online]. Available at: http://www.pubpat.org/pubpat-and-aclu-ask-supreme-court-to-rule-that-patents-on-breast-cancer-genes-are-invalid.htm [accessed: 1 February 2013].

56 *Mayo Collaborative Services* v. *Prometheus Laboratories Inc.* 132 S. Ct. 1289 (2012).

57 In *Laboratory Corp. of America Holdings v. Metabolite Laboratories Inc.* 126 S. Ct. 2921 (2006), the majority of five judges of the Supreme Court of the United States ruled that the writ of certiorari had been improvidently granted, and dismissed the action. Breyer J wrote a dissenting judgment, with the support of Stevens J and Souter J. See Rimmer, above n 26, 110-137.

58 *Mayo Collaborative Services* v. *Prometheus Laboratories Inc.* 132 S. Ct. 1289 at 1297 (2012).

59 Ibid., 1301.

60 Ibid.

grant of patents that tie up their use will inhibit future innovation' – particularly where 'patented process amounts to no more than an instruction to "apply the natural law"'.[61] In his reasoning, Breyer J was, in particular, influenced by the work of Professors Mark Lemley and Rebecca Eisenberg on patentable subject matter.[62]

Summing up, Breyer J considered the competing policy arguments by the parties and the amici curiae in the case. His Honour counterpointed the views. On side, 'Prometheus, supported by several *amici*, argues that a principle of law denying patent coverage here will interfere significantly with the ability of medical researchers to make valuable discoveries, particularly in the area of diagnostic research.'[63] On the other side,

> the American Medical Association, the American College of Medical Genetics, the American Hospital Association, the American Society of Human Genetics, the Association of American Medical Colleges, the Association for Molecular Pathology, and other medical organizations tell us that if 'claims to exclusive rights over the body's natural responses to illness and medical treatment are permitted to stand, the result will be a vast thicket of exclusive rights over the use of critical scientific data that must remain widely available if physicians are to provide sound medical care'.[64]

Considering the various policy arguments about the impact of patent law, Breyer J commented:

> 'In consequence, we must hesitate before departing from established general legal rules lest a new protective rule that seems to suit the needs of one field produce unforeseen results in another.'[65] He concluded 'that the patent claims at issue here effectively claim the underlying laws of nature themselves', and 'the claims are consequently invalid'.[66]

As a result of this decision, the Supreme Court of the United States remanded the Myriad Genetics' gene patent case to the United States Court of Appeals for the

61 Ibid.

62 Lemley, M., Risch, M., Sichelman, T. and Polk Wagner, R. 2011. Life after Bilski. *Stanford Law Review*, 63(6), 1315-1347; Eisenberg, R. 2012. Wisdom of the ages or dead-hand control? Patentable subject matter for diagnostic methods after *In re Bilski. Case Western Reserve Journal of Law, Technology & the Internet*, 3(1), 1.

63 *Mayo Collaborative Services* v. *Prometheus Laboratories Inc.* 132 S. Ct. 1289, 1304 (2012).

64 Ibid., 1304-1305.

65 Ibid., 1305.

66 Ibid.

Federal Circuit for further consideration in light of the ruling in the *Prometheus* case.[67]

The United States Court of Appeals for the Federal Circuit's Reconsideration

In the decision in *Association for Molecular Pathology* v. *USPTO and Myriad Genetics Inc.*, the United States Court of Appeals for the Federal Circuit reconsidered its ruling in light of the new *Prometheus* precedent.[68] The three judges became even more entrenched in their respective positions.

For the majority, Lourie J summarized the new decision:

> On the threshold issue of jurisdiction, we affirm the district court's decision to exercise declaratory judgment jurisdiction because we conclude that at least one plaintiff, Dr. Harry Ostrer, has standing to challenge the validity of Myriad's patents. On the merits, we reverse the district court's decision that Myriad's composition claims to 'isolated' DNA molecules cover patent-ineligible products of nature under §101 because each of the claimed molecules represents a nonnaturally occurring composition of matter. We also reverse the district court's decision that Myriad's method claim to screening potential cancer therapeutics via changes in cell growth rates of transformed cells is directed to a patent-ineligible scientific principle. We affirm the court's decision, however, that Myriad's method claims directed to 'comparing' or 'analyzing' DNA sequences are patent ineligible; such claims include no transformative steps and cover only patent-ineligible abstract, mental steps.[69]

In the judgment, Lourie J emphasized that the isolated DNA molecules were not 'products of nature':

> The isolated DNA molecules before us are not found in nature. They are obtained in the laboratory and are man-made, the product of human ingenuity. While they are prepared from products of nature, so is every other composition of matter. All new chemical or biological molecules, whether made by synthesis or decomposition, are made from natural materials. For example, virtually every medicine utilized by today's medical practitioners, and every manufactured plastic product, is either synthesized from natural materials (most often petroleum fractions) or derived from natural plant materials. But, as such, they

67 *Association for Molecular Pathology* v. *USPTO and Myriad Genetics Inc.* 132 S. Ct. 1794 (2012).

68 *Association for Molecular Pathology* v. *USPTO and Myriad Genetics Inc.* F. 3d, 2012 WL 3518509 (Fed. Cir., 2012).

69 Ibid., *1.

> are different from natural materials, even if they are ultimately derived from them. The same is true of isolated DNA molecules.[70]

Considering the composition claims, Lourie J noted: 'The remand of this case for reconsideration in light of *Mayo* might suggest, as Plaintiffs and certain amici state, that the composition claims are mere reflections of a law of nature.'[71]

Lourie responded: 'Respectfully, they are not, any more than any product of man reflects and is consistent with a law of nature.'[72] Lourie J concluded that, while 'Everything and everyone comes from nature', 'the compositions here are not natural products'.[73] Lourie J held that the inventions 'are the products of man, albeit following, as all materials do, laws of nature'.[74]

The judge emphasized that such a position was supported by past precedent, including rulings in respect of patent law and agriculture: 'The Supreme Court has repeatedly stated that changes to longstanding practice should come from Congress, not the courts.'[75] The judge noted:

> In *J.E.M. Ag Supply, Inc. v. Pioneer Hi–Bred International, Inc.*,[76] the Court rejected the argument that plants did not fall within the scope of § 101, relying in part on the fact that 'the PTO has assigned utility patents for plants for at least 16 years and there has been no indication from either Congress or agencies with expertise that such coverage is inconsistent with [federal law].'[77]

Concurring, in part, Moore J considered the doctrine of products of nature under patent law. The judge observed: 'The analysis is relatively simple if the invention previously existed in nature exactly as claimed.'[78] Interestingly, the judge discusses in this context the example of bioprospecting, mentioning that the discovery of a wild plant would be considered to be a product of nature and not subject to patent protection: 'For example, naturally existing minerals, a plant found in the wild, and physical laws such as gravity or $E=mc^2$ are not patentable subject matter, even if they were "discovered" by an enterprising inventor.'[79]

Moore J comments: 'Even if the invention was based on nature, and resulted in a living organism, it may fall within the scope of section 101.'[80] Referring to

70 Ibid., *15.
71 Ibid., *21.
72 Ibid.
73 Ibid.
74 Ibid.
75 Ibid., *22.
76 *J.E.M. Ag Supply* v. *Pioneer Hi-Bred International Inc* 534 US 124 (2001).
77 *Association for Molecular Pathology* v. *USPTO and Myriad Genetics Inc.* F. 3d, 2012 WL 3518509, *22 (Fed. Cir., 2012).
78 Ibid., *28.
79 Ibid.
80 Ibid.

plant breeding, the judge notes 'For example, "the work of the plant breeder 'in aid of nature' was patentable invention" because "a plant discovery resulting from cultivation is unique, isolated, and is not repeated by nature, nor can it be reproduced by nature unaided by man".'[81] Moore J comments more broadly that there should be judicial deference shown towards the role of the United States Congress:

> The judiciary cannot engage in an *ad hoc* innovation-based analysis, which is why the exceptions to patentability apply only to the clearest cases: a new mineral discovered in the earth, or a new plant found in the wild, or $E=mc^2$, or the law of gravity. It is Congress, with 'the constitutional authority and the institutional ability to accommodate fully the varied permutations of competing interests that are inevitably implicated by such new technology,' *Sony Corp. of America v. Universal City Studios, Inc.*, 464 U.S. 417, 431, 104 S.Ct. 774, 78 L.Ed.2d 574 (1984), who must decide whether it is necessary to change the scope of section 101 to exclude the kind of isolated DNA claims at issue here. It is not clear to me that *Chakrabarty*, *Funk Brothers*, or *Prometheus* leads inexorably to the conclusion that isolated DNA molecules are not patentable subject matter. I decline the invitation to broaden the law of nature exception. Given the complicated technology and conflicting incentives at issue here, any change must come from Congress.[82]

The judge was concerned about disrupting established business expectations: 'The settled expectations of the biotechnology industry – not to mention the thousands of issued patents –cannot be taken lightly and deserve deference.'[83] The judge observed: 'I believe leaving intact the settled expectations of property owners is particularly important in light of the large number of property rights involved, both to isolated DNA and to purified natural products generally.'[84] Moore J insisted: 'Given the complicated technology and conflicting incentives at issue here, any change must come from Congress.'[85]

Bryson J dissented from 'the court's holding that Myriad's BRCA gene claims and its claims to gene fragments are patent-eligible'.[86] The judge held:

> Although my colleagues believe our analysis of the legal question in this case should be influenced by purported expectations of the inventing community based on the PTO's past practice of issuing patents on human genes, that is in effect to give the PTO lawmaking authority that Congress has not accorded it.

81 Ibid.
82 Ibid., *36.
83 Ibid., *34.
84 Ibid.
85 Ibid., *36.
86 Ibid., *38.

> There is no collective right of adverse possession to intellectual property and we should not create one. Our role is to interpret the law that Congress has written in accordance with the governing precedents.[87]

Citing the *Prometheus* case as instructive, the judge noted: 'Just as a patent involving a law of nature must have an "inventive concept" that does "significantly more than simply describe … natural relations," *Mayo*, a patent involving a product of nature should have an inventive concept that involves more than merely incidental changes to the naturally occurring product.'[88]

The Supreme Court of the United States

Dissatisfied with this ruling, the Public Patent Foundation and the American Civil Liberties Union are appealing, once more, to the Supreme Court of the United States.[89] Daniel Ravicher, executive director of PUBPAT and co-counsel in the lawsuit, observed:

> Our genes were given to us by our creator, and not inventions created by Myriad. The patenting of human genes, even in isolated form, has the effect of giving ownership over human genes to corporations, allowing them to decide who is allowed to know what they have inside their own body.[90]

Chris Hansen, staff attorney with the ACLU Speech, Privacy and Technology Project, commented: 'In our view, the court of appeals did not fully consider or correctly apply the Supreme Court's most recent and relevant patent law decisions.'[91] The lawyer argued: 'DNA occurs naturally in the human body and cannot be patented by a single company that can then use its patents to limit scientific research and the free exchange of ideas.'[92]

The Supreme Court of the United States granted the petition to hear the appeal on 30 November 2012 – although it limited the appeal to the first question presented

87 Ibid., *47.

88 Ibid., *44.

89 American Civil Liberties Union and Public Patent Foundation 2012. *ACLU and PUBPAT Ask Supreme Court to Rule that Patents on Breast Cancer Genes are Invalid.* [Online]. Available at: http://www.aclu.org/womens-rights/aclu-and-pubpat-ask-supreme-court-rule-patents-breast-cancer-genes-are-invalid [accessed: 1 February 2013].

90 Ibid.

91 Ibid.

92 Ibid.

by the petition.[93] On 24 January 2013, the American Civil Liberties Union and the Public Patent Foundation filed the brief for the petitioners.[94] The brief argued:

> Human genes are, of course, products of nature. Genetic variants, including mutations, are also products of nature. How genes work and whether variants are harmful or not are laws of nature. Respondents Myriad Genetics *et al.* did not invent any genes or variants or cause their significance, but they did obtain patent claims on two naturally-occurring human genes known as BRCA1 and BRCA2 (so named because one of the diseases to which the genes are linked is breast cancer). The claims are not limited to any form, variation, or structure of the BRCA1 or BRCA2 genes, and they cover the BRCA genes of every person in the United States, even genes that Myriad has never seen.[95]

The brief urged the Supreme Court of the United States: 'This Court should apply long-standing doctrine without regard to PTO deference and industry reliance and find the challenged claims unpatentable.'[96]

The Supreme Court of the United States ruling in the matter of *Association for Molecular Pathology* v *Myriad Genetics Inc.* promises to be a landmark decision in respect of patent law and biotechnology.

II. The Seed Wars

The late Professor Keith Aoki[97] observed in his book *Seed Wars*: 'Contemporary debate over the legal treatment of plant genetic resources has been described as the "Seed Wars".'[98] He noted that 'the phrase "Seed Wars" as it is used here comes from a *Wall Street Journal* report that refers to the conflict implicating "access to, control over, and preservation of plant genetic resources"'.[99]

---

93 *Association for Molecular Pathology* v. *Myriad Genetics Inc.* 2012 and 2013, Proceedings and Orders, SCOTUSBlog. [Online]. Available at: http://www.scotusblog.com/case-files/cases/association-for-molecular-pathology-v-myriad-genetics-inc [accessed: 1 February 2013].

94 American Civil Liberties Union and the Public Patent Foundation. 2013. *Brief for Petitioners in the Association for Molecular Pathology v. Myriad Genetics Inc in the Supreme Court of the United States*. [Online]. Available at: http://www.americanbar.org/content/dam/aba/publications/supreme_court_preview/briefs-v2/12-398_pet.authcheckdam.pdf [accessed: 1 February 2013].

95 Ibid., 2.

96 Ibid., 55.

97 Chon, M. 2012. Law professor as artist: Themes and variations in Keith Aoki's intellectual property scholarship. *Oregon Law Review*, 90(5), 1251-1264.

98 Aoki, K. 2008. *Seed Wars: Controversies and Cases on Plant Genetic Resources and Intellectual Property*. Durham, NC: Carolina Academic Press, xiii.

99 Ibid., xiii.

There have been a number of significant skirmishes, conflicts and flash-points in the 'Seed Wars'. The Supreme Court of the United States considered whether plants were patentable subject matter in the case of *JEM Ag Supply* v. *Pioneer Hibred International*.[100] In the matter of *in re Fisher*, the United States Court of Appeals for the Federal Circuit considered the application of the utility standards to patent applications in respect of express sequence tags for maize.[101] The Supreme Court of the United States is also hearing an appeal in *Bowman* v. *Monsanto* on matters of patent exhaustion.[102] There has been disquiet over Monsanto's patent applications in relation to the enhancement of meat, including Omega-3 fatty acids.[103] The Supreme Court of Canada considered questions of patentable subject matter, patent infringement, farmers' rights and innocent bystanders, and remedies in the epic case of *Monsanto* v *Schmeiser*.[104] The European Court of Justice in *Monsanto* v *Cefetra* considered complex questions in the export of genetically modified seed to Argentina.[105] The High Court of Australia heard a constitutional challenge to the plant breeders' rights regime in *Grain Pool of Western Australia* v. *the Commonwealth*, and lower Australian courts have considered conflicts over plant breeders' rights and Franklin Barley.[106] There have also been larger concerns about the relationship between intellectual property, climate change, agriculture and food security.[107]

The Public Patent Foundation has joined the 'Seed Wars'. In March 2011, the Public Patent Foundation filed a lawsuit on behalf of 60 farmers, seeds businesses and organic agricultural organizations, challenging the validity of Monsanto's

100 *J.E.M. Ag Supply* v. *Pioneer Hi-Bred International Inc* 534 US 124 (2001).

101 *In re Fisher* 421 F.3d 1365 (C.A. Fed., 2005). For a discussion, see Rimmer, M. 2007. The new conquistadors: Patent law and expressed sequence tags. *Journal of Law, Information, and Science, 16(1),* 10-50.

102 *Bowman* v. *Monsanto* (2011), Petition to the Supreme Court of the United States, No. 11-796. [Online]. Available at: http://www.supremecourt.gov/Search.aspx?FileName=/docketfiles/11-796.htm [accessed: 1 February 2013].

103 Rimmer, M. 2011. Owning Omega-3: Monsanto and the invention of meat. *Farm Policy Journal,* 8(1), 11-21.

104 *Monsanto Canada Inc. v. Schmeiser* (2004) SCC 34; 2004 SCC 34.

105 *Monsanto Technology LLC* v. *Cefetra BV and others*, No. C-428/08, European Court of Justice (6 July 2010). For a commentary see Sanderson, J. 2012. Reconsidering plant variety rights in the EU after *Monsanto* v. *Cefetra BV*. *European Intellectual Property Review*, 34(6), 387-394.

106 *Grain Pool of Western Australia* v. *Commonwealth* (2000) 46 IPR 515; *Cultivaust* v. *Grain Pool Pty Ltd* [2004] FCA 638; *Cultivaust v. Grain Pool Pty Ltd* [2005] FCAFC 223.

107 Rimmer, M. 'Climate ready crops: intellectual property, agriculture, and climate change' and 'The Doomsday Vault: Seed banks, food security, and climate change', in Rimmer and McLennan, above n 26, 320-360 and 361-391 respectively.

patents in respect of genetically modified seeds.[108] In the press release about the case, Daniel Ravicher discussed the significance of the litigation:

> This case asks whether Monsanto has the right to sue organic farmers for patent infringement if Monsanto's transgenic seed should land on their property. It seems quite perverse that an organic farmer contaminated by transgenic seed could be accused of patent infringement, but Monsanto has made such accusations before and is notorious for having sued hundreds of farmers for patent infringement, so we had to act to protect the interests of our clients.[109]

The Public Patent Foundation sought a declaration that, if organic farmers are ever contaminated by Monsanto's genetically modified seed, they need not fear legal action for patent infringement.

Much like they did in the lawsuit against Myriad Genetics, the Public Patent Foundation relied upon testimonials from the plaintiffs affected by the patents in respect of genetically modified seed. Jim Gerritsen, a family farmer in Maine who raises organic seed, and President of lead plaintiff Organic Seed Growers and Trade Association, said: 'Today is Independence Day for America'.[110] He commented that 'Americans have the right to choice in the marketplace – to decide what kind of food they will feed their families – and we are taking this action on their behalf to protect that right to choose.'[111] Gerritsen stressed that: 'Organic farmers have the right to raise our organic crops for our families and our customers on our farms without the threat of invasion by Monsanto's genetic contamination and without harassment by a reckless polluter.'[112] Rose Marie Burroughs of plaintiff California Cloverleaf Farms said, 'We must protect the environment, farmers' livelihood, public health and people's right to non GMO food contamination.'[113]

The Public Patent Foundation Complaint

Sabrina Hassan, a legal attorney with the Public Patent Foundation, comments: 'While the Myriad Genetics matter is about patenting the laws and products

108 The Public Patent Foundation. 2011. *Organic Farmers and Seed Sellers Sue Monsanto to Protect Themselves from Patents on Genetically Modified Seed: Preemptive Action Seeks Ruling that would Prohibit Monsanto from Suing Organic Farmers and Seed Growers if Contaminated by Roundup Ready Seed*, Press Release. [Online]. Available at: http://www.pubpat.org/osgatavmonsantofiled.htm [accessed: 1 February 2013].

109 Ibid.

110 Ibid.

111 Ibid.

112 Ibid.

113 Ibid.

of nature, the Monsanto matter is about patenting products that destroy nature, biodiversity, and the environment'.[114]

In the complaint on behalf of the Organic Seed Growers and Trade Association, the Public Patent Foundation articulates the nature of the conflict in dramatic terms:

> Society stands on the precipice of forever being bound to transgenic agriculture and transgenic food. Coexistence between transgenic seed and organic seed is impossible because transgenic seed contaminates and eventually overcomes organic seed. History has already shown this, as soon after transgenic seed for canola was introduced, organic canola became virtually extinct as a result of transgenic seed contamination. Organic corn, soybean, cotton, sugar beet and alfalfa now face the same fate, as transgenic seed has been released for each of those crops, too. And transgenic seed is being developed for many other crops, thus putting the future of all food, and indeed all agriculture, at stake.[115]

The complaint notes that the 'Plaintiffs are largely organic farmers and organic seed businesses, but also include non-organic farmers who nonetheless wish to farm without transgenic seed.'[116]

The Public Patent Foundation stresses that 'Monsanto is a chemical company that was previously responsible for introducing to the world Agent Orange, DDT, PCB's and other toxins.'[117] Monsanto is now the world's leading proponent of transgenic seed and holds many patents relating thereto that it has aggressively asserted against literally hundreds of farmers, including those farmers who became contaminated by Monsanto's transgenic seed through no fault of their own.'[118]

The Public Patent Foundation contends that Monsanto's patents are invalid, because among other things, they fail to satisfy the constitutional requirement that inventions promote 'the Progress of Science' and the threshold patent demand of 'utility':

> The principal one of which is that patents on transgenic seed fail to satisfy the requirement of both the Constitution and the Patent Act that only technology with a beneficial societal use may be patented. U. S. Const., Art. I, §8, cl. 8 ('To promote the Progress of Science and *useful* Arts') (emphasis added); 35 U.S.C.

---

114 Rimmer, M. 2011. Skype interview with Sabrina Hassan of the Public Patent Foundation, Canberra-New York, 18 October 2011.

115 The Public Patent Foundation. 2012. *Organic Seed Growers and Trade Association et al. v. Monsanto Company and Monsanto Technology LLC (2012) No 11-cv-2163-NRB. First Amended Complaint.* [Online]. Available at: http://www.pubpat.org/assets/files/seed/OSGATA-v-Monsanto-Complaint.pdf [accessed: 1 February 2013].

116 Ibid.

117 Ibid.

118 Ibid.

> §101 ('Whoever invents or discovers any new and *useful* process, machine, manufacture, or composition of matter, or any new and *useful* improvement thereof, may obtain a patent therefor') (emphasis added). As Justice Story wrote in 1817, to be patentable, an invention must not be 'injurious to the wellbeing, good policy, or sound morals of society,' and 'a new invention to poison people ... is not a patentable invention'. Because transgenic seed, and in particular Monsanto's transgenic seed, is 'injurious to the wellbeing, good policy, or sound morals of society' and threatens to 'poison people,' Monsanto's transgenic seed patents are all invalid.[119]

This complaint has echoes of other case law. In the area of copyright law, there were constitutional challenges to the copyright term extension – some emphasizing that such a lengthening of term did not promote 'the Progress of Science and the Useful Arts'.[120] In the area of patent law, Monsanto faced a challenge in respect of the utility of its express sequence tags. The notion of utility and morality has been explored in contemporary times in the context of some of the test cases involving animal-human chimera.[121]

The Public Patent Foundation[122] draws upon the 1817 authority of Justice Story in *Lowell v. Lewis*, in which the judge held that inventions that are 'injurious to the well-being, good policy, or sound morals of society' are unpatentable.[123] The Foundation seizes upon his examples of unpatentable inventions, including 'a new invention to poison people, or to promote debauchery, or to facilitate private assassination'.

Furthermore, the Public Patent Foundation contend that 'Monsanto's patents are additionally invalid for other failures to meet the requirements of patent law, including that each violates the prohibition against double patenting, each is anticipated or rendered obvious by prior art, and each fails to satisfy the requirements of written description, enablement and best mode.'[124] The Foundation elaborates:

> Monsanto began applying for patents on glyphosate tolerance in the mid 1980s. Its first patents on the trait were granted in 1990 and are now expired. After pursuing its earliest patents on glyphosate resistance, Monsanto continued to seek and receive patents on Roundup Ready technology for over two decades. Although the United States patent system allows improvements on existing inventions, it does not permit a party to extend its monopoly over a field of

119 Ibid.

120 See Rimmer, M. 2007. *Digital Copyright and the Consumer Revolution: Hands off my iPod*. Cheltenham and Northampton, MA: Edward Elgar.

121 See Rimmer, above n 26, 98-103.

122 The Public Patent Foundation, above n 115.

123 *Lowell v. Lewis*, 15 F. Cas. 1018 at 1019 (C.C.D. Mass. 1817).

124 The Public Patent Foundation, above n 115.

> invention by receiving a patent that expires later than and is not patentably distinct from a patent it already owns. In acquiring the transgenic seed patents, Monsanto unjustly extended its period of patent exclusivity by duplicating its ownership of a field of invention already covered by other Monsanto patents. Monsanto's transgenic seed patents are thus invalid for violating the prohibition against double patenting. Monsanto's patents are invalid because prior art exists that anticipates or renders obvious each of their claims. Monsanto's patents are invalid for failure to meet the written description, enablement and best mode requirements.[125]

The Foundation is contending that Monsanto has been engaging in the practice of 'evergreening' – seeking to obtain a longer period of protection that warranted under patent law through strategic means.

The Public Patent Foundation also argues that

> Monsanto's patents would also not be infringed by Plaintiffs because, amongst other things, Plaintiffs do not intend to use Monsanto's transgenic seed, any seed possessed by Plaintiffs that may be contaminated by Monsanto's transgenic seed is not covered by any valid and properly construed claim of any patent in suit, and Monsanto's patents rights in transgenic seed exhaust upon the authorized distribution by Monsanto to its customers.[126]

The Foundation maintains that 'Transgenic seed that may come to contaminate Plaintiffs cannot infringe any of Monsanto's transgenic seed patents.'[127] The Foundation argues that the 'Plaintiffs cannot be held to have infringed any Monsanto transgenic seed patent if Plaintiffs become contaminated by Monsanto's transgenic seed through no intentional act of their own.'[128] The Foundation contends: 'Because Monsanto sells, licenses and distributes its transgenic seed in a manner such that contamination of Plaintiffs is reasonably foreseeable, any patent rights Monsanto may have in transgenic seed are exhausted when it or its distributors make authorized sales to customers.'[129] In this context, it will be interesting to see whether the United States courts take any notice of the Supreme Court of Canada ruling on Percy Schmeiser and patent infringement in respect of Monsanto's patents on genetically modified canola.

Furthermore, the Public Patent Foundation submitted that 'Monsanto's patents are also unenforceable because, among other things, Monsanto has committed

125 Ibid.
126 Ibid.
127 Ibid.
128 Ibid.
129 Ibid.

misuse, Monsanto is equitably estopped from enforcing them, and Monsanto commits trespass when its transgenic seed contaminates another.'[130]

The Public Patent Foundation also maintained that, in any case,

> Monsanto would not be entitled to any remedy under law or equity even if its patents were held to be valid, infringed and enforceable against Plaintiffs, as no economic injury happens to Monsanto and the public interest would not support granting Monsanto an injunction when its patented seed contaminates another.[131]

The Farm and Ranch Freedom Alliance

There was an amicus curiae brief filed in support of organic farms by a collection of groups including the Farm and Ranch Freedom Alliance, Slow Food USA, Virginia Independent Consumers and Farmers Association, and others.[132] The organizations included 'members who feed grain to livestock and poultry, who certify organic production of crops, who use grains or cotton as raw ingredients for other products, and who consume or use products made from these crops'.[133] A number of the 'amici are in privity with the affected farmers, either by buying their crops or otherwise conducting business with them such as through organic certification of the crops'.[134] In its statement of interests, the *amici curiae* emphasized that the case had larger implications for a wide range of farmers, food producers, and retailers: .

> While the Plaintiffs are at the most immediate risk of suit for patent infringement by Monsanto, the legal principles involved in this Court's decision will have even broader ramifications. For example, livestock and poultry farmers who feed grain to their animals face issues of GM-contaminated feed. Organic certifiers must make decisions about whether or not, and under what conditions, to require testing for GM contamination as part of the certification process. Food processors, whether they operate on a large-scale or simply bake a few loaves of bread for a local farmers market, use ingredients that may be contaminated with Monsanto's patented products. And, ultimately, almost every American consumer somehow makes use products made from corn, soybeans, canola, sugar beets, or cotton, all of which may implicate the scope and enforceability

130 Ibid.

131 Ibid.

132 Farm and Ranch Freedom Alliance and others. 2011. *An Amicus Brief in Support of Plaintiffs' Opposition to Defendants' Motion to Dismiss in Organic Seed Growers and Trade Association v. Monsanto Company and Monsanto Technology LLC.* [Online]. Available at: http://farmandranchfreedom.org/sff/Amicus-brief-filed.pdf [accessed: 1 February 2013].

133 Ibid.

134 Ibid.

> of Monsanto's patents. The entire food chain is impacted by the spread of Monsanto's patented crops.[135]

The amici curiae stressed that they had an 'interest in having the scope and enforceability of Monsanto's patents determined by this Court, both because of the direct issue of their own liability and because of the indirect impact the decision will have on the availability of these crops'.[136] The amici curiae emphasized that the outcome of this case 'will have repercussions for almost every American'.[137]

Summarizing its argument, the Farm and Ranch Freedom Alliance and their fellow organizations contended that farmers and other participants in the agricultural industry were frightened by the spectre of patent infringement lawsuits: 'Because of the nature of Monsanto's patented seeds, the individual Plaintiffs and the farmer members of Plaintiff organizations (hereinafter collectively "Plaintiff farmers") cannot avoid infringing on Monsanto's patents unless they entirely abandon growing corn, soybeans, canola, cotton, sugar beets, and, as of this year, alfalfa.'[138] The brief asserted: 'Not only does Monsanto's patented technology inevitably lead to infringement through no fault of the Plaintiffs, but, by their design, the majority of Monsanto's patented crops only provide the alleged benefits if a farmer applies herbicides, specifically Roundup®, directly to the crop.'[139] The brief worried: 'Due to Monsanto's decision to release patented seeds and market them for widespread planting, it is now impossible for farmers to remain 100% free of genetically modified crops because of the multitude of ways that contamination can occur.'[140] The brief maintained: 'The constant threat of a patent infringement suit by Monsanto creates significant, unquantifiable costs for the Plaintiff farmers and similarly situated farmers.'[141] In its submission, the Farm and Ranch Freedom Alliance was concerned about the dangers of patent infringement in respect of self-replicating GM seeds: 'While the fault for such GM contamination lies with Monsanto's decision to create and market this self-replicating, uncontrollable technology, the patent doctrine of strict liability means that the farmers still face liability.'[142] The Alliance observes: 'The biological reality is that a farmer who raises corn, cotton, soybeans, canola, sugar beets, or alfalfa cannot reliably avoid infringing on Monsanto's patent.'[143] The group maintained that farmers faced a Hobbesian choice: 'The farmer is left with a choice: risk infringement or refrain

135 Ibid., 1.
136 Ibid.
137 Ibid.
138 Ibid., 5.
139 Ibid.
140 Ibid., 5-6.
141 Ibid., 6.
142 Ibid., 14.
143 Ibid., 17.

from raising these crops at all.'[144] There are echoes here of the arguments about patent infringement in the Supreme Court of Canada dispute between Monsanto and Percy Schmeiser.[145]

In its analysis, the Farm and Ranch Freedom Alliance focuses upon the technology of genetically modified crops as being self-replicating:

> Monsanto seeks to reap all of the benefits of a patented seed – particularly the necessity for farmers to buy their products year after year indefinitely, since new seed is needed every single year – but to take no responsibility for the reality that its technology, by its very nature, induces others to infringe.[146]

The Alliance alleged: 'By patenting a self-replicating product, one virulent in its spread, Monsanto has created a situation in which it can pick and choose targets for enforcement activity.'[147] The Alliance maintained: 'Monsanto has, by its own admission, implemented an aggressive campaign to enforce its patent rights.'[148] It noted that 'According to Monsanto's website, in the section addressing lawsuits against farmers for patent infringement, the company has filed 128 lawsuits and settled "almost 700 matters" out of court.'[149] The Alliance maintains: 'By patenting a self-replicating product, one virulent in its spread, Monsanto has created a situation in which infringement is a certainty and occurs at ever-increasing levels.'[150] The Alliance argues that Monsanto's pledge not to sue for 'trace contamination' is neither meaningful or enforceable: 'By deliberately keeping their pledge vague, Monsanto perversely expects human error and biological realities to compound the amount of contamination to the point where seed contamination is substantial and near-universal, making infringement unavoidable.'[151]

The Farm and Ranch Freedom Alliance also notes that Monsanto's claim that it has never sued an organic producer does not protect the plaintiff farmers. The Alliance comments that organic farmers are answerable to the managers of certification schemes for organic food:

> While the organic regulations technically require only that certified organic farmers not knowingly plant GM seed, certifiers may spot-check crops and, if GM contamination is found, require the farmer to take steps to reduce contamination. Moreover, certified organic farmers face significant economic losses if they cannot sell their crops as certified organic, or have to pull land

144 Ibid., 17-18.
145 *Monsanto Canada Inc. v. Schmeiser*, [2004] 1 SCR 902, 2004 SCC 34.
146 Farm and Ranch Freedom Alliance and others, above n 132, 6.
147 Ibid., 7.
148 Ibid.
149 Ibid.
150 Ibid., 11.
151 Ibid.

> out of the program. So they would have a potentially expensive counter-claim against Monsanto for the loss of organic markets from contamination, making it logical that Monsanto has so far hesitated to sue a certified organic producer.[152]

The Farm and Ranch Freedom Alliance commented: 'Many people are not interested in including Monsanto's products in the food chain'.[153] In conclusion, the Farm and Ranch Freedom Alliance summed up its argument, stressing: 'It is critical for both these farmers and for all the parties involved in the food chain, including the Amici, that this Court consider this case and provide a clear declaration of their rights.'[154]

The United States District Court for the Southern District of New York

The New York Times provided a vivid portrait of the hearing before the United States District Court.[155] The journalist Julia Moskin observed of the crowd of spectators, listening to the hearing:

> Silent in flannel shirts and ponytails, farmers from Saskatchewan and South Dakota, Mississippi and Massachusetts lined the walls of a packed federal courtroom in Manhattan last week, as their lawyers told a judge that they were no longer able to keep genetically modified crops from their fields. The hearing is part of a debate that is coming to life around the country, in courtrooms and Occupy sites, in boardrooms and online, with new petitions, ballot initiatives and lawsuits from California to Maine.[156]

She commented that 'the point made by the suit is that, according to the regulations that govern American agriculture, it's these unwilling farmers who must prevent Monsanto's products from trespassing onto their land'.[157]

In 2012 of the *Organic Seed Growers and Trade Association* v. *Monsanto Company and Monsanto Technology LLC*, Buchwald J of the United States District Court for the Southern District of New York was unsympathetic to the action brought by the organic farmers, holding that the action failed to satisfy the case or controversy requirement.[158] The judge depicted the issue in these terms:

---

152 Ibid., 12.

153 Ibid., 15.

154 Ibid., 18.

155 Moskin, J. 2012. Modified crops tap a wellspring of protest. *The New York Times*. [Online, 7 February]. Available at: http://www.nytimes.com/2012/02/08/dining/a-suit-airs-debate-on-organic-vs-modified-crops.html?_r=4 [accessed: 1 February 2013].

156 Ibid.

157 Ibid.

158 *Organic Seed Growers and Trade Association* v. *Monsanto Company and Monsanto Technology LLC* 851 F. Supp. 2d 544 (2012).

> Growers who wish to use defendants' seeds must obtain limited-use licenses to do so. Because subsequent generations of plants grown from these seeds will also contain the glyphosate-tolerance trait, these licenses authorize growers to use the seed only to grow a single crop; growers are not authorized to harvest and plant the second-generation seed produced from the original planting, or to sell seeds containing the patented trait outside authorized channels of distribution. Despite these restrictions, some unlicensed – and unintended – use of transgenic seeds is inevitable. Like any other seeds, transgenic seeds may contaminate non-transgenic crops through a variety of means, including seed drift or scatter, crosspollination, and commingling via tainted equipment during harvest or post-harvest activities, processing, transportation, and storage.[159]

Buchwald J observed that 'No plaintiffs claim that contamination has yet occurred in any crops they have grown or seed they have sold.'[160] The judge noted that 'one plaintiff farmer claims that seed he considered purchasing in 2010 was contaminated with genetically engineered seed', and one plaintiff seed distributor claims that it received shipments contaminated with genetically modified seed in each of 2005, 2008, 2009, and 2010' – but 'Neither plaintiff asserts that the offending seeds were covered by defendants' patents.'[161]

Buchwald J portrayed the potential of organic farmers infringing patents relating to genetically modified crops as a theoretical possibility. The judge noted: 'Contamination can theoretically affect non-transgenic farmers by lowering the price for which their crops may be sold and potentially resulting in rejected shipments.'[162] Buchwald J commented:

> Additionally, inadvertent growth of crops with patented traits may potentially subject a farmer to liability for patent infringement. While defendants investigate hundreds of possible patent infringers each year, between 1997 and April 2010 they filed just 144 lawsuits to enforce their patent rights against farmers. Defendants, moreover, have never filed a patent-infringement suit against a certified organic farm or handling operation over the presence of patented traits in its operations, and they stated at oral argument that they have never sued a party who did not 'want to make use of the traits that are manifested in [defendants'] transgenic products'. Indeed, defendants have expressly declared that it is not their policy 'to exercise [their] patent rights where trace amounts of our seed or traits are present in [a] farmer's fields as a result of inadvertent means'. Nevertheless, plaintiffs allege without specification that defendants have accused certain non-intentional users of Monsanto's seed of patent

159 Ibid., 548.
160 Ibid.
161 Ibid.
162 Ibid., 549.

> infringement and threatened them with litigation. No plaintiffs claim to have been so threatened.[163]

Buchwald J observed: 'While there is no evidence in the record that any farmer has ever been decertified as organic by the U.S. Department of Agriculture National Organic Program (the "NOP") because of seed contamination, we do not foreclose that hypothetical possibility.'[164] There have certainly, though, been instances of decertification of organic produce because of commingling genetically modified crops in other jurisdictions – such as Australia.[165]

Buchwald J noted that the 'Plaintiffs do not allege that defendants have ever demanded royalty payments from plaintiffs, identified any of plaintiffs' conduct as potentially infringing, or even initiated any contact with plaintiffs whatsoever.'[166] The judge was of the view that

> instead, plaintiffs posit the existence of an actual case or controversy based on: (1) defendants' pattern of enforcing their patent rights against non-plaintiff farmers through litigation or threats of litigation; (2) plaintiffs' assertion of the 'implicit threat' in defendants' statement that it is not their policy to enforce their patent rights against farmers whose crops inadvertently acquire trace amounts of patented seeds or traits; and (3) defendants' refusal to provide plaintiffs with a blanket covenant not to sue.[167]

Summing up, Buchwald J maintained that, in all the circumstances, there was not a case of controversy:

> Defendants' patent-infringement suits against other, dissimilar parties cannot by themselves create subject matter jurisdiction, and the diaphanous allegations that defendants have threatened but not sued unintentional infringers do not add much weight to the substantiality of the dispute. Nor have plaintiffs pointed to any other circumstances that bolster the objective reasonableness of their claims of threat of injury. Defendants' statement regarding the exercise of their patent rights against inadvertent infringers is, if anything, a source of comfort rather than worry.[168]

---

163 Ibid.

164 Ibid.

165 See the dispute in the Supreme Court of Western Australia, Stephen William Marsh, 'Writ of Summons', Submission in *Marsh v. Baxter*, CIV 1561 of 2012, 3 April 2012.

166 *Organic Seed Growers and Trade Association* v. *Monsanto Company and Monsanto Technology LLC* 851 F. Supp. 2d 544, 551 (2012).

167 Ibid.

168 Ibid., 556.

The judge held: 'Taken together, it is clear that these circumstances do not amount to a substantial controversy and that there has been no injury traceable to defendants.'[169]

The decision of Buchwald J is a poor one – especially in light of the litigation over Myriad Genetics' patents in respect of genetic testing for breast cancer and ovarian cancer, where it was found that there was a case or a controversy in that dispute.[170]

Appeal and Intellectual Property Professors' Amicus Brief

The Public Patent Foundation is appealing against the ruling of Buchwald J to the United States Court of Appeals for the Federal Circuit.[171] In short, the Foundation argues that 'The district court's requirement that Monsanto be directly enforcing its patents against Plaintiffs before they can seek redress of the injury being caused to them by those patents is contrary to the law.'[172]

In July 2012, a number of intellectual property professors filed a brief to the United States Court of Appeal for the Federal Circuit, maintaining that the decision of the United States District Court should be reversed.[173] The academics included Professors Margo A. Bagley, Michael J. Burstein, Rochelle C. Dreyfuss, Brett Frischmann, Erika R. George, Shubha Ghosh, Megan M. La Belle, Kali Murray, Ted Sichelman, Katherine J. Strandburg and Melissa Wasserman. The brief emphasized: 'Although we have diverse perspectives on the substance of patent law, we are uniformly concerned about the development of the law of Article III standing as it relates to causes of action challenging the validity of issued patents.'[174] The law professors maintain:

> Finding standing in such circumstances will not open the floodgates of litigation. The conditions described above do not authorize citizen suits to challenge patents. And to the extent the Court remains concerned about abusive or opportunistic litigation, it has a variety of prudential and discretionary doctrines

169 Ibid.

170 *Association for Molecular Pathology* v. *USPTO and Myriad Genetics Inc.* 702 F. Supp. 2d 181 (S.D.N.Y., 2010).

171 The Public Patent Foundation. 2012. *Brief of Appellants to the United States Court of Appeals for the Federal Circuit in Organic Seed Growers and Trade Association v. Monsanto Company and Monsanto Technology LLC.* [Online]. Available at: http://www.pubpat.org/assets/files/seed/OrganicSeedCAFCBrief.pdf [accessed: 1 February 2013].

172 Ibid., 36.

173 Intellectual Property Professors. 2012. *Amici Curiae Brief of Intellectual Property Professors in Support of Plaintiffs-Appellants and Supporting Reversal in Organic Seed Growers and Trade Association v. Monsanto Company and Monsanto Technology LLC.* [Online]. Available at: http://www.pubpat.org/assets/files/seed/OrganicSeedCAFCAmiciProfessors.pdf [accessed: 1 February 2013].

174 Ibid., 1.

> that it can use to tailor its jurisdiction to the circumstances of particular cases without erecting a constitutional bar to standing that applies across a variety of litigation postures. To the contrary, broad standing to challenge the validity of patents ensures that the courts can effectively play their critical role in screening out invalid patents.[175]

The law professors insist that 'Injury in fact is present when a plaintiff has been deterred from engaging in present economic activity by the threat of a potentially invalid patent.'[176] The academics note that 'The risk that innovative activity may run afoul of a patent may often be present long before infringement is imminent; yet that risk is a present injury when it adversely affects investment behaviour.'[177] Citing the case of *Bresnick v. U.S. Vitamin Corp.*,[178] the academics observed that 'Judge Learned Hand recognized long ago that a patent can serve as a "scarecrow," deterring activity by its mere presence in a field of endeavour.'[179]

The law professors comment that it is critically important that parties should be able to access the courts to challenge the validity of patents:

> Access to the courts to test the validity of issued patents is particularly important because of the administrative structure of the patent system. The Patent and Trademark Office ('PTO') receives over 500,000 utility patent applications each year and issues over 200,000 patents annually. Because the vast majority of patents are never litigated or licensed, the PTO must necessarily devote fewer resources to examination than parties expend in contested litigation over patent validity. The courts therefore provide an essential forum for litigants to test the validity of economically significant patents. Indeed, empirical evidence suggests that 30 to 50 per cent of validity challenges that are litigated to judgment result in a finding of invalidity. Broad standing rules ensure that the courts remain widely accessible as a check on patent quality.[180]

The law professors observed: 'Accordingly, this Court should hold that plaintiffs who allege that they have been restrained from engaging in economic activity due to the presence of a potentially invalid patent have standing to sue for a declaration that the patent is invalid.'[181]

175 Ibid., 3.
176 Ibid., 2.
177 Ibid.
178 *Bresnick v. US Vitamin Corp.*, 139 F.2d 239, 242 (2d Cir., 1943).
179 Intellectual Property Professors, above n 173.
180 Ibid., 17-18.
181 Ibid., 20.

In January 2013, the Public Patent Foundation engaged in oral argument to ask the Court of Appeals for the Federal Circuit to reinstate the case.[182] Jim Gerritsen, an organic potato farmer from Maine and President of OSGATA, the lead plaintiff in the lawsuit, commented on the case:

> Family farmers need and deserve the right to farm. We have a right to grow good food and good seed for our families and our communities without the threat of trespass and intimidation. We need Court protection so that our families will be able to carry on our farming tradition and help keep America strong.[183]

Arguably, the case of the organic farmers should be reinstated by the United States Court of Appeals for the Federal Circuit, and considered properly by the lower court.

Comparative Law

The patent dispute between the organic farmers and the biotechnology company, Monsanto, is of wider international significance. There have been parallel disputes between organic farmers, and biotechnology companies in other jurisdictions.

In Australia, there has been a lively policy debate about legal liability in relation to genetic modified crops. There has been debate about breaking the moratoria on the commercial release of genetically modified crops in Victoria.[184] In Western Australia, there has been a dramatic legal conflict between an organic farmer, Stephen Marsh, and Michael Baxter, a farmer of genetically modified canola.[185] In April 2012, Marsh filed a lawsuit against Baxter in the Supreme Court of Western Australia, alleging negligence and nuisance for causing the presence of GM plants and seeds on Marsh's property.[186] Marsh is concerned that he has lost

---

182 Public Patent Foundation. 2013. *Farmers Join Hundreds to Protest Outside Monsanto Hearing and Demand End to Monsanto's Campaign of Intimidation in Rural America Over GMOs*. [Online]. Available at: http://www.woodprairie.com/pressrelease_011013 [accessed: 1 February 2013].

183 Ibid.

184 Carrick, D. 2005. GM contamination of crops, *The Law Report, ABC*. [Online]. Available at: http://www.abc.net.au/radionational/programs/lawreport/gm-contamination-of-crops/3373262 [accessed: 1 February 2013].

185 For media commentary on the dispute, see Hobday, L. 2011. Farmer sues over GM contamination, *PM, ABC*. [Online]. Available at: http://www.abc.net.au/pm/content/2011/s3280227.htm [accessed: 1 February 2013]; Norman, J. 2011. Farmer sues over GM contamination, *News, ABC*. [Online]. Available at: http://www.abc.net.au/news/2011-07-28/farmers-sues-neighbour-over-gm-contamination/2813046 [accessed: 1 February 2013].

186 Stephen William Marsh, 'Writ of Summons', Submission in *Marsh v. Baxter*, CIV 1561 of 2012, 3 April 2012: for a commentary see Lee, J. *Cultivating Coexistence: Regulating GM Crops and Organic Farms*. Honours Thesis, ANU College of Law, 4 June 2012.

certification as an organic farmer, because of the alleged contamination, and has lost the premium price for his organic produce. Marsh has sought an injunction to stop the nuisance; and otherwise, damages with interest and costs.

Similarly, in Canada, Larry Hoffman and farmers from Saskatoon brought an action against Monsanto and Bayer Crop Science.[187] The action was driven by the Saskatchewan Organic Directorate-Organic Agriculture Protection Fund. Cameron JA noted: 'The action was ostensibly taken on behalf of numerous organic grain farmers for the recovery of damages alleged to have been suffered by them as a result of the introduction by the appellants of strains of genetically modified canola for use by farmers generally.'[188] Cameron JA observed:

> The statement of claim alleges that the appellants and other organic grain farmers suffered financial losses as a result of the introduction and commercial use of Roundup Ready and Liberty Link canola. More particularly, the statement of claim alleges that these strains of genetically modified canola, which are open-pollinating, inevitably find their way onto their fields, thus preventing them from producing and marketing organically grown canola, and putting them to extra expense in producing other organically grown crops. The statement of claim goes on to allege that the companies are liable for these losses on the bases of negligence, nuisance, and trespass. In addition, the statement of claim alleges that the companies are liable on the bases of *The Environmental Management and Protection Act*, S.S.1983-84, c. E-10.2, and *The Environmental Assessment Act*, S.S. 1979-80, c. E-10.1… In consequence, the companies are alleged to be liable for the losses sustained by the appellants, and liable to declaratory relief declaring their conduct unlawful. [189]

At first instance, and appeal, the Canadian courts held that this action did not merit certification as a class action.

The Canadian Professor Jeremy de Beer has argued that such Canadian cases have failed adequately to address the issue of the liability of biotechnology companies to farmers harmed by the spread of genetic modified seeds.[190] He contends that, while nuisance and negligence may be appropriate in certain circumstances, trespass law perhaps is the best cause of action to deal with the problem. Jeremy de Beer, though, notes that trespass law will need be reformed and modernized to address the challenges posed by biotechnology. He has called for the creation of 'biotrespass' as a viable cause of action to deal with the problem.

---

187 *Hoffman* v. *Monsanto Canada Inc.*, 2007 SKCA 47 (CanLII).

188 Ibid.

189 Ibid.

190 de Beer, J. 2007. Biotrespass. *Bulletin of Science, Technology and Society*, 7(4), 287-300.

III. America Invents

In addition to its role in patent disputes, the Public Patent Foundation has also played an active role in the policy debate over patent law reform:

> PUBPAT advocates for improvements to the patent system and educates the public about the esoteric issues implicated by patents, specifically how they impact everyday life. PUBPAT advocates on behalf of sound patent policy before the courts, in the PTO, and through participation in relevant conferences, while PUBPAT educates through public presentations, one-on-one discussions and publication of articles. PUBPAT also uses its other activities as the basis for normative research studying the patent system to determine whether it strikes the proper balance between incentivizing innovation and benefiting the public.[191]

The Public Patent Foundation has been particularly involved in debates over the *Leahy-Smith America Invents Act* 2011 (US), patent quality, patent infringement, patent exceptions and patent remedies. The organization has also displayed an interest in questions about trade mark law, consumer rights and labelling – particularly in the context of food labelling.

The Leahy-Smith American Invents Act 2011 (US)

In the United States, there has been halting efforts to reform the patent administration and system.

In his first term, President Barack Obama hailed the *Leahy-Smith America Invents Act* 2011 (US) as a piece of landmark patent law reform.[192] In September 2011, Obama discussed the new patent regime at Thomas Jefferson in Alexandria, Virginia.[193] He commented that the new legal infrastructure would help promote entrepreneurial activity and innovation in the digital economy:

> When Thomas Edison filed his patent for the phonograph, his application was approved in just seven weeks. And these days, that process is taking an average of three years. Over the last decade, patent applications have nearly tripled. And because the Patent Office doesn't have the resources to deal with all of them, right now there are about 700,000 applications that haven't even been opened yet. These are jobs and businesses of the future just waiting to be created. The CEOs who are represented here today, all of them are running companies

---

191 The Public Patent Foundation 2012. *PUBPAT Activities*. [Online]. Available at: http://www.pubpat.org/activities.htm [accessed: 1 February 2013].

192 *Leahy-Smith America Invents Act*, 2011 (US) Pub. L. No. 112-29.

193 President Barack Obama. 2012. *Remarks at the Signing of the America Invents Act, The White House*. [Online]. Available at: http://www.whitehouse.gov/the-press-office/2011/09/16/remarks-president-signing-america-invents-act [accessed: 1 February 2013].

> that were based on creativity and invention and the ability to commercialize good ideas. And somewhere in that stack of applications could be the next technological breakthrough, the next miracle drug, the next idea that will launch the next Fortune 500 company. And somewhere in this country – maybe in this room – is the next Thomas Edison or Steve Jobs, just waiting for a chance to turn their idea into a new, thriving business.[194]

Obama comments that the *Leahy-Smith America Invents Act* 'reforms the outdated patent process' and 'cuts away the red tape that slows down our inventors and entrepreneurs'. [195] He insisted that the legislation 'that will put a dent in the huge stack of patent applications waiting for review' and 'will help startups and small business owners turn their ideas into products three times faster than they can today'.[196] Moreover, Obama believed that *Leahy-Smith America Invents Act* 'will improve patent quality and help give entrepreneurs the protection and the confidence they need to attract investment, to grow their businesses, and to hire more workers'.[197] Obama comments that patent should system should encourage and support a dynamic, innovative economy in the United States: 'This is the economy we need to build – one where innovation is encouraged, education is a national mission, and new jobs and businesses take root right here in America.'[198]

Unconvinced by such rhetoric, the Public Patent Foundation has been a provocative critic of the *Leahy-Smith America Invents Act*, arguing that the legislative reforms fail to properly address underlying systematic issues with patent law in the United States. In a piece for the *Huffington Post*, Daniel Ravicher questions the confidence of Obama that the patent regime will be a boon for United States small businesses:

> The America Invents Act will indeed most likely result in the Patent Office issuing more patents (it already issues about 4,500 per week), but one has to ask if that's really a good thing, and in particular if it's a good thing for small businesses here in America, as President Obama proclaims. My answer to that question is that the president has a fundamentally naive understanding of the patent system today, because getting more patents of their own will not help American small businesses if everyone else is getting more patents, too. In reality, an enlarged patent system with more patents being given out by the Patent Office gives a huge relative advantage to larger competitors who have teams of patent lawyers that can file dozens of patent applications for every one

194 Ibid.
195 Ibid.
196 Ibid.
197 Ibid.
198 Ibid.

> filed by a small business. The phrase 'Don't take knives to a gun fight.' comes to mind. [199]

In this context, Ravicher contends that 'more government regulation is bad, not good, especially for small businesses, and patents are just another form of government regulation that interferes with free markets and competition'.[200] While abhorring patent abolitionism, Ravicher argues: 'We do need to fix our patent system so that it rewards innovation, not manipulation.'[201] He comments: 'Unfortunately, the America Invents Act will not do that. It may actually result in making things worse.'[202]

The Public Patent Foundation has been somewhat concerned about aspects of the *Leahy-Smith America Invents Act* in relation to the false marking of patent claims.[203] In November 2011, the Public Patent Foundation launched a challenge against a provision of the *America Invents Act*, which retroactively denies standing for any false marking complainant who cannot prove a competitive injury.[204] The Public Patent Foundation comments that such a retroactive denial of standing violates due process rights affected by the Fifth Amendment:

> Retroactive congressional action that deprives a private party of its property violates the Due Process Clause of the Fifth Amendment if it is not 'supported by a legitimate legislative purpose.' Further, the retroactive effect of legislation must separately have a legitimate legislative purpose apart from the prospective substantive changes ... All PUBPAT suggests is that retroactive statutes can violate the Due Process Clause and that the America Invents Act's retroactive application of substantive changes to the false marking statute in a way that does nothing but deprive PUBPAT of its property interest in order to bestow a private benefit on McNeil is an example of precisely such a violation.[205]

The Public Patent Foundation sees this action as part of its push against 'the negative effects that over-patenting, unmerited patenting and excessive patent rights can have on society'.[206] However, some commentators have been sceptical

---

199 Ravicher, D. 2012. Is Obama's patent reform hurting small business? *Huffington Post Small Business.* [Online, 5 October]. Available at: http://www.huffingtonpost.com/daniel-b-ravicher/patent-reform_b_1503763.html [accessed: 1 February 2013].

200 Ibid.

201 Ibid.

202 Ibid.

203 The Public Patent Foundation. 2011. Fighting the retroactive elimination of false marking claims. *Patent Law Blog, Patently-O.* [Online]. Available at: http://www.patentlyo.com/patent/2011/11/the-public-patent-foundation-pubpat-has-continued-to-push-against-the-negative-effects-that-over-patenting-unmerited-pate.html [accessed: 1 February 2013].

204 Ibid.

205 Ibid.

206 Ibid.

of the prospects of success of such a constitutional challenge. Brian Pandya, an intellectual property litigator, responded to the dispute thus: 'In this author's view, the amendments plainly do not violate the due process rights of *qui tam* plaintiffs such as PUBPAT.'[207] He maintains that 'more fundamentally, contrary to PUBPAT's argument, there was nothing arbitrary or illegitimate in Congress' decision to amend Section 292'.[208] Pandya's response does have some resonance. It is true that the Supreme Court of the United States have been unsympathetic to constitutional challenges to intellectual property laws. Nonetheless, in terms of patent law and policy, there is a need to have broad standing to address issues such as the false marking of patents.

Patent Quality

The Public Patent Foundation has been active in seeking to improve patent quality – both by pushing for tougher patent threshold standards, and by pushing for greater rigour in patent examination.

The Public Patent Foundation filed an amicus brief in the dispute before the Supreme Court of the United States in *KSR International Co.* v *Teleflex Inc.*[209] The Foundation emphasized that it had 'an interest in ensuring that obviousness is maintained as a bar to patentability and that the Federal Circuit's teaching-suggestion-motivation requirement for the combination of known technologies to be obvious is rejected'. The Public Patent Foundation emphasized:

> The Federal Circuit's 'teaching-suggestion-motivation' test contradicts both the Patent Act and this Court's precedent. The test modifies the Patent Act and therefore is an improper judicial encroachment on the legislative powers of Congress. The Federal Circuit's test requiring that the exact same combination be explicitly suggested previously in order to prove 'obviousness' grants unjustified rewards for non-innovative subject matter and prevents competitors from using common knowledge that should be freely available to all.[210]

---

207 Pandya, B. 2011. Last gasp for false patent marking cases: The Public Patent Foundation challenges the constitutionality of the America Invents Act. *Washington Legal Foundation's The Legal Pulse*. [Online, 14 November]. Available at: http://www.wileyrein.com/publications.cfm?sp=articles&id=7641 [accessed: 1 February 2013].

208 Ibid.

209 Public Patent Foundation. 2006. *Brief of AARP, Patients not Patents Inc., and the Public Foundation as Amici Curiae in Support of Petitioner in KSR v. Teleflex*. [Online]. Available at: http://www.pubpat.org/assets/files/AmicusBriefs/PUBPAT_KSR_SCt_Brief.pdf [accessed: 1 February 2013].

210 Ibid.

The Supreme Court of the United States was sensitive to such concerns in its ruling in the 2007 case of *KSR International Co.* v *Teleflex Inc.*[211] Kennedy J commented: 'We build and create, by bringing to the tangible and palpable reality around us new works based on instinct, simple logic, ordinary inferences, extraordinary ideas and sometimes even genius.'[212] Kennedy J emphasized: 'As progress beginning from higher levels of achievement is expected in the normal course, the results of ordinary innovation are not the subject of exclusive rights under the patent laws.'[213] He added: 'Were it otherwise patents might stifle, rather than promote, the progress of useful arts.'[214]

Daniel Ravicher of the Public Patent Foundation despairs of the low patent quality in the current regime:

> The Supreme Court has continually acknowledged the importance of maintaining high patent quality, saying just a few years ago that, 'the results of ordinary innovation are not the subject of exclusive rights under the patent laws [because w]ere it otherwise patents might stifle, rather than promote, the progress of useful arts'. Unfortunately, the American patent system today is suffering from extremely low patent quality. Every Tuesday the Patent Office issues 4,500 patents after spending, on average, less than a couple of days in reviewing the merits of each patent application. When asked to reconsider the merits of patents it previously issued, the Patent Office concedes that the vast majority of them have questionable validity. When patents end up in court as a result of patent owners suing alleged infringers, a large percentage of the time those asserted patents are found to have been improperly granted. The result is a polluted patent system littered with trash patents that impede technological development.[215]

He maintains that it is 'critically important to the success of our patent system that it maintain high patent quality and ensures only deserving patents are issued'.[216]

In this context, Peter Drahos has made a number of recommendations on how to improve the patent system in his book, *The Global Governance of Knowledge: Patent Offices and their Clients*.[217] He observes that 'patent offices are important

211 *KSR International Co.* v. *Teleflex Inc.* 550 US 398 (2007).

212 Ibid., 427.

213 Ibid.

214 Ibid.

215 Ravicher, D. 2012. The patent pollution problem: Its causes, effects and solutions. *The Huffington Post*. [Online, 1 May]. Available at: http://www.huffingtonpost.com/daniel-b-ravicher/the-patent-pollution-problem_b_1465478.html [accessed: 1 February 2013].

216 Ibid.

217 Drahos, P. 2010. *The Global Governance of Knowledge: Patent Offices and their Clients*. Cambridge: Cambridge University Press.

nodal centres, but they are centres that are integrated into a private governance network'.[218] Drahos maintains that

> the challenge for states is to use their sovereignty over patent administration to make patent offices part of public nodal governance networks, so that patent offices regulate in ways that complement rather than undermine the public policy goals of states in areas such as health, renewable energy, and information technology.[219]

Patent Infringement

The Public Patent Foundation has been interested in the larger public policy debate about patent infringement – particularly as it pertains to biotechnology.

The issue will be considered by the Supreme Court of the United States in *Bowman* v. *Monsanto Company and Monsanto Technology LLC*.[220]

At first instance, in *Monsanto Company and Monsanto Technology LLC* v. *Bowman*, Young J of the United States District Court for the Southern District of Indiana considered an action brought by the biotechnology company against a farmer, alleging that he had infringed patents for genetically modified soybean seeds.[221] The judge granted the owner's motion for summary judgment of patent infringement and entered judgment for $84,456.20. The farmer appealed.

In the 2011 appeal, in *Monsanto Company and Monsanto Technology LLC* v *Bowman*, the United States Court of Appeals for the Federal Circuit affirmed the ruling.[222] Linn J noted that 'This case presents the court with another question of patent infringement by farmers planting the progeny of genetically altered seeds covered by U.S. patents'.[223] The farmer argued that Monsanto's patent rights had been exhausted with respect to all the Roundup Ready soybean seeds present in grain elevators. Linn J rejected the patent exhaustion argument, ruling:

> [P]atent exhaustion does not bar an infringement action. Even if Monsanto's patent rights in the commodity seeds are exhausted, such a conclusion would be of no consequence because once a grower, like Bowman, plants the

---

218 Ibid., 327.

219 Ibid., 327-328.

220 *Bowman* v. *Monsanto Company and Monsanto Technology LLC* (2011), Petition to the Supreme Court of the United States, No. 11-796. [Online]. Available at: http://www.supremecourt.gov/Search.aspx?FileName=/docketfiles/11-796.htm [accessed: 1 February 2013].

221 *Monsanto Company and Monsanto Technology LLC v. Bowman* 686 F. Supp. 2d 834 (2009).

222 *Monsanto Company and Monsanto Technology LLC v. Bowman* 657 F. 3d 1341 (2011).

223 Ibid., 1343.

> commodity seeds containing Monsanto's Roundup Ready® technology and the next generation of seed develops, the grower has created a newly infringing article. 'The fact that a patented technology can replicate itself does not give a purchaser the right to use replicated copies of the technology. Applying the first sale doctrine to subsequent generations of self-replicating technology would eviscerate the rights of the patent holder.' The right to use 'do[es] not include the right to construct an essentially new article on the template of the original, for the right to make the article remains with the patentee.' The court disagrees with Bowman that a seed 'substantially embodies' all later generation seeds, at least with respect to the commodity seeds, because nothing in the record indicates that the 'only reasonable and intended use' of commodity seeds is for replanting them to create new seeds. Indeed, there are various uses for commodity seeds, including use as feed. While farmers, like Bowman, may have the right to use commodity seeds as feed, or for any other conceivable use, they cannot 'replicate' Monsanto's patented technology by planting it in the ground to create newly infringing genetic material, seeds, and plants. [224]

The United States Court of Appeals for the Federal Circuit affirmed 'the district court's holding that patent exhaustion does not apply to Bowman's accused second-crop plantings'.[225] The Supreme Court of the United States is hearing an appeal in the matter of *Bowman* v. *Monsanto Company and Monsanto Technology LLC*.[226] The question presented is: 'Whether the Federal Circuit erred by (1) refusing to find patent exhaustion in patented seeds even after an authorized sale and by (2) creating an exception to the doctrine of patent exhaustion for self-replicating technologies?'[227]

The Public Patent Foundation has filed an amicus brief to the Supreme Court of the United States.[228] Summarizing its argument, the Public Patent Foundation emphasized that:

> The doctrine of exhaustion provides a necessary limit on the ability of patentees to pervert markets beyond the scope of their patents. Sophisticated patent owners like Monsanto can be expected to take limits on their patent rights, including

224 Ibid., 1347-1348.

225 Ibid., 1349.

226 *Bowman* v. *Monsanto Company and Monsanto Technology LLC* (2011), Petition to the Supreme Court of the United States, No. 11-796. [Online]. Available at: http://www.supremecourt.gov/Search.aspx?FileName=/docketfiles/11-796.htm [accessed: 1 February 2013].

227 Ibid.

228 Public Patent Foundation. 2012. Brief as *Amicus Curiae* in Support of the Petitioner in *Bowman* v. *Monsanto Company* et al. in the Supreme Court of the United States, 10 December. [Online]. Available at http://www.americanbar.org/content/dam/aba/publications/supreme_court_preview/briefs-v2/11-796_pet_amcu_ppf.authcheckdam.pdf [accessed: 1 February 2013].

> exhaustion, into account when designing their business models. Monsanto's attempt to contract around exhaustion is impotent and not legally sustainable. Patent rights are not recreated anew upon the production of new seed from purchased seed in which Monsanto's patent rights were exhausted upon the authorized sale.[229]

A number of other *amicus curiae* briefs have been submitted by other parties – including civil society groups, economists, law professors, associations, and representatives of agricultural and biotechnology industries.[230]

The decision of the Supreme Court of Canada in *Monsanto Canada Inc.* v. *Schmeiser* also provides a thoughtful reflection on patent infringement and agricultural biotechnology.[231]

Patent Defences, Exceptions and Limitations

Arguably, the Public Patent Foundation could play a more active role in the public policy debate about patent defences, exceptions and limitations.

In his policy guide for the Quaker UN Office and the International Centre for Trade and Sustainable Development, Carlos Correa considers the relationship between intellectual property and food security.[232] Recognizing that 'Achieving food security is a major objective of the international community', Correa comments: 'The effort to reduce the number of people suffering from food insecurity faces many obstacles, including underinvestment in agriculture and rural development, substantial increases in food prices and poverty.'[233] He observes:

> A number of factors related to the production of seeds and control over plant materials and technologies can also impair attainment of food security. The field related to the development and production of commercial seeds is highly concentrated. There is a growing orientation of research and development (R&D) toward commercially attractive crops and varieties, while traditional seed supply systems are eroded. The spread of commercial plant varieties has significantly reduced the genetic variability and increased the vulnerability of agricultural production. In addition … some forms of intellectual property rights

229 Ibid., 2.

230 *Bowman v. Monsanto Co.*, Proceedings and Orders, SCOTUSBlog. [Online]. Available at: http://www.scotusblog.com/case-files/cases/bowman-v-monsanto-co/ [accessed: 1 February 2013].

231 *Monsanto Canada Inc. v. Schmeiser* (2004) SCC 34; 2004 SCC 34.

232 Correa, C. 2012. *TRIPS-Related Patent Flexibilities and Food Security: Options for Reform*, Geneva: Quaker UN Office and the International Centre for Trade and Sustainable Development. [Online]. Available at: http://ictsd.org/i/publications/146808/ [accessed: 1 February 2013].

233 Ibid., 2.

> (IPRs) can create barriers to the free flow of information and materials (e.g. through farmers' exchange of seeds), which is essential to sustained levels of food production as well as development of new varieties.[234]

Correa comments that

> achieving food security requires a legal system that preserves a free flow of knowledge and plant materials, so as to permit a continuous process of improvement and adaptation of plant varieties through research and breeding, the availability of a diversity of materials to respond to local needs and changing conditions (including those generated by climate change), and the diversification (as opposed to concentration) of the supply of technologies and seeds.[235]

In his work, *TRIPS-Related Patent Flexibilities and Food Security: Options for Developing Countries*, Carlos Correa makes a number of recommendations in respect of patent exceptions in relation to agriculture.

First, Correa contends that there is a need for an innocent bystanders' defence, given that 'the presence of a trait in a plant protected by a patent may or may not be intentional, as a patented gene trait may disseminate by natural means and appear in plantations unintentionally'.[236] Referring to the dispute in *Monsanto Canada Inc.* v. *Schmeiser*,[237] Correa recommends that 'national laws should exempt from liability unintentional infringement caused by the dissemination of patented genetic materials'.[238] Such an option would also be a means of addressing the concerns of organic farmers – such as those represented by the Public Patent Foundation – about patent infringement resulting from unintentional use of patented plants.

Second, Correa contends that patent law should clarify the position of farmers' rights, much like the plant breeders' rights regime. He observes that some countries have clarified the position of the treatment of patent rights in respect of reproductive materials:

> French law, for instance, makes it clear that plant material can be multiplied or reproduced where it has been legally put on the market by the patent holder or with his or her consent, where this was the purpose for which the material has been marketed; the obtained material, however, cannot be subsequently used for further reproduction or multiplication.[239]

---

234 Ibid., 2.
235 Ibid., 3.
236 Ibid., 13.
237 *Monsanto Canada Inc. v. Schmeiser* (2004) SCC 34; 2004 SCC 34.
238 Correa, above n 232, 13.
239 Ibid., 13.

Correa observes that a farmers' privilege to save and reuse seeds under patent law would be a 'key component of a legal regime sensitive to food security policies, since it would reduce costs of production and promote the diversification of the sources of supply of seeds'.[240] He recommends that 'national patent laws should, where plants and/or their components are patentable, introduce exceptions equivalent to the farmers' privilege under Plant Variety Protection'.[241] Again, the policy option of a farmers' privilege would be of benefit to the organic farmers represented by the Public Patent Foundation – it would help alleviate concerns about farmers being sued for patent infringement in respect of saved seed.

Third, Correa discusses intellectual property exceptions in respect of research and breeding relating to agriculture. He notes that a breeders' exemption is a mandatory feature under the UPOV Convention, and 'optimizes variety improvement by ensuring that germplasm sources remain accessible to all the community of breeders'.[242] Correa recommends:

> The continuous improvement of plant varieties requires freedom to undertake research and breeding where patented materials are involved. Exceptions to this effect should be adopted even where a country opts not to grant patents on plants (or plant varieties). Even if it opts not do so, such exceptions should be included if the patentability of plant components is permitted. National laws may permit the commercialization of the newly obtained varieties, on the basis of non-remunerative exceptions.[243]

He notes, though, that 'the compatibility of an exception under patent law – equivalent in its scope and effects to the breeders' exception – with the TRIPS Agreement (articles 28 and 30) has not been tested yet'.[244]

A related issue is the defence of experimental use. The defence of experimental use has been interpreted very narrowly by the United States courts in a number of matters.[245] The Supreme Court of the United States, though, has taken a broader view of the defence safe harbour in relation to regulatory activities for pharmaceutical drugs.[246] Other jurisdictions have sought to engage in patent law reform in respect of patent exceptions. After a long policy debate,[247] the Australian Government established a statutory defence of experimental use under the *Intellectual Property Laws Amendment (Raising the Bar) Act* 2012 (Cth). There is

---

240 Ibid.

241 Ibid.

242 Ibid., 14. Correa is quoting from the UPOV Council at this point.

243 Ibid.

244 Ibid.

245 *Madey v. Duke University* 307 F. 3d 1351 (2002).

246 *Merck KGaA* v. *Integra Lifesciences I, Ltd* 545 U.S. 193 (2005).

247 Rimmer, M. 2005. The freedom to tinker: Patent law and experimental use. *Expert Opinion on Therapeutic Patents*, 15(2), 167-200.

a need to provide a broad and flexible research exemption to protect experimental researchers from the predatory behaviour of 'patent trolls'.

Finally, Correa also broaches the question of compulsory licensing. He suggests that 'national patent laws should provide for compulsory licenses in cases where the exploitation of a protected plant variety would infringe a patent'.[248] He argues that 'a cross-license may be granted to the patent owner with regard to the concerned variety'.[249]

Patent Remedies

There has been much concern about the remedies awarded in respect of matters of patent infringement. In the 2012 case of *Monsanto Co.* v *E.I. du Pont de Nemours & Co*, a Federal Court jury found in favour of Monsanto in a patent case, holding that DuPont and DuPont Pioneer had engaged in willful infringement of its patent rights.[250] The jury awarded damages of $1 Billion on Monsanto. The finding of willful infringement could lead to an increased award of damages. David Snively, Monsanto's Executive Vice President and General Counsel, was elated by this ruling:

> Importantly, this verdict highlights that all companies that make early and substantial investments in developing cutting edge technology will have their intellectual property rights upheld and fairly valued. This verdict also underscores that DuPont's unauthorized use of the Roundup Ready technology was both deliberate and aimed at rescuing its own failed technology. The materials uncovered from DuPont files during this case highlight that DuPont's senior leaders were actively working to hide the fact their OGAT technology had failed and were using elaborate schemes to cover that up with the unlicensed use of our technology. They knew the OGAT technology didn't work for years, but opted to tell a much different story to their customers and to Wall Street. It is deeply disappointing that repeated requests to DuPont's leadership and board to investigate their own internal actions were not addressed and corrected, which ultimately required the matter go to trial.[251]

---

248 Correa, above n 232.

249 Ibid., 15.

250 *Monsanto Co. v. E.I. du Pont de Nemours & Co.*, 09cv686, U.S. District Court, Eastern District of Missouri (St. Louis) 2012.

251 Monsanto Co. 2012. *Federal Court Jury Finds in Favor of Monsanto in Patent Case, Du Pont and DuPont Pioneer Found Guilty of 'Willful' Infringement of Intellectual Property*. [Online]. Available at: http://monsanto.mediaroom.com/2012-08-01-Federal-Court-Jury-Finds-In-Favor-Of-Monsanto-In-Patent-Case-DuPont-And-DuPont-Pioneer-Found-Guilty-Of-Willful-Infringement-Of-Intellectual-Property [accessed: 1 February 2013].

The company emphasized that the 'verdict represents the third time that court proceedings have determined that DuPont or DuPont Pioneer has violated a trait licensing agreement with Monsanto'.[252]

DuPont has strongly disagreed with this verdict, and has said that it will appeal against the decision.[253] DuPont Senior Vice President and General Counsel Thomas L. Sager has commented:

> DuPont believes that the evidence presented during the trial demonstrated clearly that Monsanto's Roundup Ready® soybean patent (RE 39,247) is invalid and unenforceable and that Monsanto intentionally deceived the USPTO on several occasions as it sought patent protection. Further, DuPont believes that the damages awarded of $1 billion are unjustified, particularly considering that Pioneer has never sold a single Optimum® GAT® seed and has no plans to do so in the future. DuPont's license to sell Roundup Ready® soybeans remains in place and is not impacted by this verdict. DuPont is and always has been committed to innovation and providing farmers with diverse technology options. We continue to stand by our position that we did not infringe the Roundup Ready® soybean patent and that the Monsanto patent is invalid. DuPont will continue to fight hard to ensure American farmers have choices and that no one company decides what is best for them.[254]

Du Pont is hoping to raise allegations about Monsanto's misconduct involving this patent, which were not tried in this case, and will be considered by another jury as part of DuPont's antitrust and patent misuse case against Monsanto in September 2013.

The remedies awarded in the field of patent disputes in respect of agriculture and biotechnology have been astonishing.

Trade Mark Law, Consumer Protection and Food Labelling

As part of their campaign on behalf of organic farmers, the Public Patent Foundation has also promoted policy initiatives in respect of food labelling – such as the Just Label It campaign.[255] The Public Patent Foundation commented:

---

252 Ibid.

253 DuPont. 2012. *DuPont will appeal verdict in Monsanto v. DuPont Patent Case*. [Online]. Available at: http://www.prnewswire.com/news-releases/statement-from-dupont-dupont-will-appeal-verdict-in-monsanto-vs-dupont-patent-case-164676366.html [accessed: 1 February 2013].

254 Ibid.

255 Just Label It. 2012. *We Have A Right to Know*. [Online]. Available at: http://justlabelit.org/ [accessed: 1 February 2013].

> PUBPAT encourages the public to not buy any products made with corn, soy, sugar, canola, cotton or alfalfa (this includes milk, as dairy cows eat alfalfa) unless you are certain it was made without any use of genetically modified seed. If you're not sure, call the manufacturer and ask. If they can't or don't give you a straight answer, then don't buy their product. The proponents of genetically modified seed have vigorously opposed labeling of genetically modified food here in America (although Europe and Asia have such labeling), so to make this effort easier on your fellow Americans, once you know whether certain products are derived from genetically modified seed or not, spread that information so others know. Increasing consumer awareness and demand for food not derived from genetically modified seed, even slightly, will increase the supply of the food we want, which will reduce prices and increase availability. If you want to purge genetically modified food from society, you can help do so every time you go to the grocery store or a restaurant. It's your money, spend it as you see fit.[256]

There has been a longstanding debate in the United States – particularly at a state level – over labelling genetically modified food.[257]

In 2012, there was an intense debate in California over Proposition 37.[258] James Wheaton, who filed the ballot initiative, refers to the Proposition as 'The California Right to Know Genetically Engineered Food Act'. Proposition 37 would require labelling of food sold to consumers made from plants or animals with genetic material changed in specified ways. Proposition 37 would also prohibit the marketing of such food, or other processed food, as 'natural'. The regime has a number of exceptions and limitations – including foods that are

> certified organic; unintentionally produced with genetically engineered material; made from animals fed or injected with genetically engineered material but not genetically engineered themselves; processed with or containing only small amounts of genetically engineered ingredients; administered for treatment of medical conditions; sold for immediate consumption such as in a restaurant; or alcoholic beverages.

The advocates of Proposition 37[259] argued that the measure would 'give us the right to know what is in the food we eat and feed to our families'.[260] In their view,

256 The Public Patent Foundation. 2012. *OSGATA v. Monsanto: Seed Patents*. [Online]. Available at: http://www.pubpat.org/monsanto-seed-patents.htm [accessed: 1 February 2013].

257 Carrick, above n 184.

258 California General Election. 2012. *Proposition 37*. [Online]. Available at: http://voterguide.sos.ca.gov/propositions/37 [accessed: 1 February 2013].

259 California Right to Know. 2012. *Label Genetically Engineered Foods*. [Online]. Available at: http://www.carighttoknow.org/ [accessed: 1 February 2013].

260 California General Election, above n 258.

the measure 'requires labelling of food produced using genetic engineering, so we can choose whether to buy those products or not'.[261] The champions of the measure maintain: 'We have a right to know.'[262] The advocates of the measure maintain:

> Fifty countries around the world – representing more than 40% of the world's population – already require GMO labeling, including all of Europe, Japan, India and China. Polls show that more than 90% of Americans want to know if their food is genetically engineered. We are free to choose what we want to eat and feed our children. The free market is supposed to provide consumers with accurate information about products so we can make informed choices.[263]

The opponents of the measure argue that 'Proposition 37 is a deceptive, deeply flawed food labeling scheme, full of special-interest exemptions and loopholes'.[264] The opponents allege that the regime would 'create new government bureaucracy costing taxpayers millions, authorize expensive shakedown lawsuits against farmers and small businesses, and increase family grocery bills by hundreds of dollars per year'.[265]

The ballot measure for Proposition 37 in California was unsuccessful in the November 2012 election. In the end, the No position received 6,356,236 votes and 51.5 per cent of the total votes, while the Yes position obtained 5,986,652 votes and 48.5 per cent of the total votes.[266] A disappointed Kristin Lynch, Pacific region director of NGO Food & Water, lamented about the loss:

> While support for genetically engineered food labels has never been stronger, the incessant drumbeat of misleading and outright false industry advertising was barely able to defeat this popular measure. While disappointed in the result, we believe that this movement to label GE foods is stronger than ever and we will continue to build a robust national grassroots campaign to push for mandatory labeling across the country.[267]

---

261 Ibid.

262 Ibid.

263 California Right to Know. 2012. *Facts – Yes on Prop 37*. [Online]. Available at: http://www.carighttoknow.org/facts [accessed: 1 February 2013].

264 California General Election, above n 258.

265 Ibid.

266 Ballotpedia. 2012. *California Proposition 37, Mandatory Labelling of Genetically Engineered Food*. [Online]. Available at: http://ballotpedia.org/wiki/index.php/California_Proposition_37,_Mandatory_Labeling_of_Genetically_Engineered_Food_(2012) [accessed: 1 February 2013].

267 Vaughan, A. 2012. Prop 37: Californian voters reject GM food labelling. *The Guardian*. [Online, 7 November]. Available at: http://www.guardian.co.uk/environment/2012/nov/07/prop-37-californian-gm-labelling [accessed: 1 February 2013].

The food labeling movement has vowed to continue its advocacy in the United States – with a proposal for a ballot initiative in Washington State in November 2013, and legislative efforts to require labeling in Connecticut and Vermont.[268]

Conclusion

This chapter has provided a critical evaluation of the model of the Public Patent Foundation, and its role in 'patent-busting'. Particular attention has been paid to two case studies – involving the battles over Myriad Genetics' patents in respect of genetic testing for BRCA1 and BRCA, and Monsanto's patents in relation to genetically modified seed.

In the United States, the Electronic Frontier Foundation has sought emulate the success of the Public Patent Foundation, with its own 'patent-busting' initiative focused on information technology patents.[269] The Electronic Frontier Foundation complain that 'every year numerous illegitimate patent applications make their way through the United States patent examination process without adequate review'.[270] The Foundation notes: 'The problem is particularly acute in the software and Internet fields where the history of prior inventions (often called "prior art") is widely distributed and poorly documented.'[271] The Foundation discusses the harm caused by bad information technology patents:

> Illegitimate patents can also threaten free expression. More and more people are using software and Internet technology to express themselves online. Website and blogging tools are increasingly popular. Video and audio streaming technology is ubiquitous. E-mail and Instant Messaging have reached users of all ages. Yet because patents can be anywhere and everywhere in these technologies, the average user has no way of knowing whether his or her tools are subject to legal threats. Patent owners who claim control over these means of community discourse can threaten anyone who uses them, even for personal non-commercial purposes. We lose much if we allow overreaching patent claims to reduce the tremendous benefits that software and technology bring to freedom of expression.[272]

---

268 Hull, D. 2012. With Proposition 37 defeated, food movement vows to fight on. *Mercury News*. [Online, 7 November]. Available at: http://www.mercurynews.com/elections/ci_21950655/proposition-37-defeated-food-movement-vows-fight [accessed: 1 February 2013].

269 Electronic Frontier Foundation. 2012. *Patent-Busting Project*. [Online]. Available at: https://www.eff.org/patent-busting [accessed: 1 February 2013].

270 Ibid.

271 Ibid.

272 Ibid.

In the style of a Wild West 'Wanted' poster, the Foundation has established a 'rogue's gallery' of objectionable information technology patents. The Foundation hopes to engage in crowd-sourcing in the search for prior art, which could invalidate the patent applications: 'Take a look at the Top Ten Most Wanted culprits on our list, and let us know if you have any leads on technology that predates them (called "prior art") that we can use to challenge their validity.'[273] The Electronic Frontier Foundation has adopted a collaborative approach in its 'patent-busting' activities – it has invited contributions from the public; it has built upon the work of the Chilling Effects Clearinghouse; and collaborated with the Internet Archive, the Public Patent Foundation and various technology law school clinics.

The civil group noted: 'Millions of people in the developing world do not have access to the medicines they need because they – or their government – cannot afford them.'[274] Médecins Sans Frontières was concerned that 'overly-strong patent protection, where patent-holders charge high prices for the drugs they produce, is a major concern for civil society groups and international organizations attempting to reach the millions of people waiting for treatment'.[275] The civil group observed that patent oppositions were important means of oversight of the patent regime: 'Asking questions or bringing an action to challenge the grant of a patent is not necessarily a bad reflection on the patent office or its staff, but part of a system of necessary checks and balances intended to protect the public interest.'[276] Médecins Sans Frontières maintained: 'This database is a tool which can be used to explore how to challenge unfair patents and their negative impact on access to medicines.'[277] This project will be used to test the validity of key patents in the fields of pharmaceutical drugs, medicines and methods of human treatment.

The model of the Public Patent Foundation should be adopted in other jurisdictions. The United States is not unique or exceptional in concerns about the quality of patents in areas of critical significance. In Australia, for example, there has been much controversy over patent law and biotechnology – particularly in respect of Genetic Technologies Limited's non-coding DNA and genomic mapping patents;[278] Myriad Genetics' patents in respect of BRCA1 and BRCA2;[279] and Woo-Suk Hwang's patent applications in respect of stem cell research.[280] A public-spirited entity liked an Australian Public Patent Foundation could play a

273 Ibid.

274 Ibid.

275 Ibid.

276 Ibid.

277 Ibid.

278 Rimmer, M. The alchemy of junk: Patent law and non-coding DNA. *The University of Ottawa Law and Technology Journal*, 3(2), 539-599.

279 *Cancer Voices Australia* et al. v. *Myriad Genetics Inc*. et al. (2010), Federal Court of Australia, Statement of Claim, 8 June.

280 Woo-Suk Hwang [2004] APO 24 (9 September 2004); Cyranoski, D. 2008. Hwang Work Patent Granted. *Nature*, 455(7213), 571.

useful role in testing the validity and the scope of controversial, high-profile patents like these ones. In the European Union, civil society groups and non-government organizations have made constructive use of opposition proceedings against patent applications. This has been evident in disputes over patent applications relating to plants and animals; Myriad Genetics' patents in respect of BRCA1 and BRCA2, controversies over stem cell patents and objections in respect of biopiracy. However, such interventions have been quite ad hoc – and dependent upon the interest of civil society groups. A European Public Patent Foundation could provide a useful professional, organized focal point for the engagement and participation of civil society and the wider public in patent matters.

The model of the Public Patent Foundation could also be adapted for other fields of intellectual property law. The field of copyright law does seem to have equivalents to the Public Patent Foundation. There is a Fair Use Project[281] under copyright law, the Chilling Effects Clearinghouse[282] and the Organization for Transformative Works, [283] to name a few defenders of the public interest under copyright law. By contrast, the registration systems of intellectual property do not have the same activity. The area of trade mark law could benefit from a publicly-spirited organization, which would test the validity of controversial trade marks. Designs law lacks a champion of the public interest. The special United States regime of plant patents could benefit from independent challenges. [284] The sui generis area of plant breeders' rights[285] would also benefit from a public ombudsman, which would test the validity of plant breeders' registrations. Such an entity might be of particular interest to those in the agricultural community – working in farming, breeding and such areas. The area of access to genetic resources[286] could also benefit from the existence of a public entity willing to challenge the validity of intellectual property rights which involved the biopiracy of genetic resources.

281 Center for Internet and Society. 2012. *The Fair Use Project.* [Online]. Available at: cyberlaw.stanford.edu/taxonomy/term/374 *[accessed: 1 February 2013].* See also Rimmer, M. 2010. A Fair Use Project for Australia: Copyright Law and Creative Freedom. *Copyright Reporter*, 28(3), 165-212.

282 Chilling Effects Clearinghouse. 2012. *Copyright.* [Online]. Available at: http://www.chillingeffects.org/copyright [accessed: 1 February 2013].

283 Organization for Transformative Works. 2012. *Organization for Transformative Works.* [Online]. Available at: http://transformativeworks.org [accessed: 1 February 2013].

284 For a history of plant patents see Pottage, A. and Sherman, B. 2010. *Figures of Invention: A History of Modern Patent Law.* Oxford: Oxford University Press, 153-182.

285 For an overview of the area see Sanderson, J. 2011. Intellectual property and plants: Constitutive, contingent and complex, in *Emerging Issues in Intellectual Property Law*, edited by Bowrey, K., Nicol, D. and Handler, M. Oxford: Oxford University Press.

286 Lawson, C. 2012. *Regulating Genetic Resources: Access and Benefit Sharing in International Law*, Cheltenham and Northampton, MA: Edward Elgar.

Index

References to illustrations are in **bold**